GLYCEMIC DENSITY

CONTINUING THE GLUCOSE REVOLUTION

BY

RICHARD A. PRICE

ISBN 0-7414-4246-9

Published by:

INFINITY PUBLISHING.COM

1094 New DeHaven Street, Suite 100
West Conshohocken, PA 19428-2713
Info@buybooksontheweb.com
www.buybooksontheweb.com
Toll-free (877) BUY BOOK
Local Phone (610) 941-9999
Fax (610) 941-9959

Printed in the United States of America

Printed on Recycled Paper

Published August 2007

Table of Contents

PART I

INTRODUCTION TO GLYCEMIC DENSITY

CHAPTER ONE: BRIEF OVERVIEW OF GLYCEMIC DENSITY

AWAKENING

What is the secret of healthy weight management? Perhaps you have struggled with this issue as I have. My wake-up call came when I developed diabetes at the age of fifty. Desiring to avoid the possible negative consequences, I became determined to learn all that I could about healthy nutrition and weight management. Controlling diabetes with diet and exercise is the best way of dealing with the disease if a person is able to do so. With this goal in view, I went through a series of steps and missteps involving the diabetic food exchanges, low carbohydrate dieting, and various other approaches. All of this culminated in the discovery of what I call Glycemic Density.

DISCOVERY OF GLYCEMIC DENSITY

Glycemic Density is built upon two foundations which are: (1) The Glycemic index and (2) Energy Density. The glycemic index measures how rapidly carbohydrate foods are digested and metabolized into glucose or blood sugar. A related concept is the glycemic load which applies the glycemic index to specific serving sizes. Energy density or calorie density is a measurement of calories per gram of food. This results in a number ranging from zero to nine. The most calorie dense nutrient is pure fat which has nine calories per gram. Pure carbohydrate or pure protein would have an energy density of four, because there are four calories per gram of carbohydrate or protein. The addition of water and/or fiber will lower the energy density further, due to the fact that they have no caloric value. Energy density allows us to gain the maximum hunger satisfaction with the least amount of calories.

I was thinking about the principle of energy density, when I wondered if there would be any value to a carbohydrate calorie density which would measure carbohydrate calories per gram of food. This would, of course, give a value that ranged from zero to four. Then a thought flashed through my mind. Why not apply the glycemic index to this carbohydrate measurement? This would measure the glycemic impact in a gram of food. Thus was the concept of Glycemic Density born.

Glycemic Density Score (GDS)
= (Net-Carbohydrate-Grams * 4
/ Total-Grams-in-Serving)
*** (Glycemic-Index / 100).**

Note: Throughout this book the terms Glycemic Density and Glycemic Density Score (GDS) will be used interchangeably.

Glycemic Density is also a number between zero and four. A Glycemic Density of zero would indicate that the food has no measurable glycemic impact. Fish, poultry, meat, cheese and eggs are in this category as well as many low carbohydrate or non-starchy vegetables. Pure glucose would have the top measure of four, because it is 100% sugar and has a glycemic index of 100%. Glycemic Density allows us to get the maximum hunger satisfaction with the least glycemic impact.

HOW THIS BOOK IS ORGANIZED

This chapter is meant to be a very brief introduction to the concept of Glycemic Density. Chapter two gives a list of foods divided into very low, low, moderate and high Glycemic Density rankings. Within each ranking, the foods are divided into food groups. Within each group, foods are then listed in Glycemic Density order. This is all the information you may need to know in order to apply this principle.

Scientific explanations are deferred to chapters three, four and five. Chapter three gives a more detailed explanation of the glycemic index. Chapter four does the same for energy density. Finally, chapter five gives a more detailed explanation of Glycemic Density. A separate section is reserved for tables which support the development and use of this principle. Some of these may prove to be very useful.

By applying the ideas discussed in this book, I was able to lose nearly eighty pounds and learn how to control my weight and blood sugar through diet and exercise. There are many men and women who could learn to do the same.

I am not a professional nutritionist. It is my hope that over time, the principle of Glycemic Density will penetrate the scientific and academic world and more accurate information will be forthcoming.

CHAPTER TWO: WHAT SHALL WE EAT?

This chapter contains a list of foods that correspond to the Glycemic Density principle that we have been discussing. This will be followed by other information and an abbreviated food list. The foods are divided into:

Very low - Glycemic Density = 0 to .20
Low - Glycemic Density = .21 to .40
Moderate - Glycemic Density = .41 to .60
High - Glycemic Density = .61 and higher

Note that a Glycemic Density of "-" means that the glycemic index was not able to be calculated. Some foods such as cottage cheese or low carbohydrate vegetables have such a low level of carbohydrates that the glycemic index is not able to be tested. People just can't eat that much for a good test. They probably do have some glycemic impact.

These rankings have yet to be scientifically evaluated. They actually compare quite favorably with rankings established for energy density. There are some foods with low Glycemic Density Scores containing fat and / or protein that may be better positioned with the energy density rankings for those who are dieting through calorie reduction. These I have highlighted and also described with a separate table.

VERY LOW GLYCEMIC DENSITY FOOD

The very low Glycemic Density list contains those foods with a Glycemic Density of no more than .20. We should select the majority of our foods from this list in order to satisfy our hunger. For all of these tables, I use GDS for Glycemic Density Score.

VERY LOW GLYCEMIC DENSITY FOOD (Cont)

Food	GDS
Low Carbohydrate Vegetables	
Spinach, cooked	-
Cauliflower, cooked	-
Lettuce, iceberg, raw	-
Mushrooms, cooked	-
Radishes, raw	-
Cucumber, raw	-
Celery, raw	-
Lettuce, Romaine, raw	-
Zucchini, cooked	-
Brussels sprouts, cooked	-
Cabbage, raw	-
Tomatoes, raw	-
Cauliflower, raw	-
Tomatoes, cooked	-
Asparagus, cooked	-
Green peppers, raw	-
Spinach, raw	-
Green beans, cooked	-
Broccoli, cooked	-
Broccoli, raw	-
Snap peas, raw, whole	-
Green onions, raw	-
Fruit	
Strawberries	.09
Grapefruit	.10
Cherries	.12
Peaches	.14

VERY LOW GLYCEMIC DENSITY FOOD (Cont)

Food	**GDS**
Fruit (Continued)	
Peaches, canned in juice	.15
Pears, canned in juice	.16
Plumbs	.16
Oranges	.17
Apples	.17
Pears	.19
Cantaloupe	.20
Applesauce, no added sugar	.20
Grains and Starchy Vegetables	
Squash, winter, cooked	-
Carrots, raw	.05
Zoom or Wheatena hot breakfast cereal	.12
Beets, canned, drained	.14
Oatmeal, old fashioned	.14
All Bran, Kellogg's, with milk	.16
Protein Foods	
Cottage cheese, low fat (1%)	**-**
Cottage cheese, reduced fat (2%)	**-**
Cottage cheese, plain	**-**
Cottage cheese, nonfat	**-**
Milk, soy	.03
Milk, whole	.05
Soybeans, fresh	**.05**
Milk, reduced fat (2%)	.06
Milk, low fat (1%)	.06
Milk, nonfat	.06
Yogurt, nonfat flavored, no sugar added	.07
Milk, chocolate, nonfat, no sugar added	.07

VERY LOW GLYCEMIC DENSITY FOOD (Cont)

Food	GDS
Protein Foods (Continued)	
Yogurt, nonfat, plain	.08
Kidney beans, canned	**.09**
Lentils, cooked	**.12**
Milk, chocolate (1%)	.14
Pinto beans, canned	**.18**
Lima beans, cooked	**.18**
Fats	
Olives	**.00**
Avocados	**.00**
Peanuts (For peanut butter, natural is best)	**.01**
Cashews (May also select other nuts)	**.06**
Indulgences	
Pudding, sugar free, nonfat milk	.12
Custard	**.19**

LOW GLYCEMIC DENSITY FOOD

The low Glycemic Density list contains those foods with a Glycemic Density of greater than .20 but no more than .40. My evaluation is that the foods with lower levels can be used to satisfy hunger, but exercise restraint as the number begins to rise. Medical science has yet to evaluate these principles.

Food	GDS
Fruit	
Watermelon	.21
Apricots	.21
Fruit cocktail, canned in juice	.23
Peaches, canned in light syrup	.23
Kiwi	.25

LOW GLYCEMIC DENSITY FOOD (Continued)

Food	GDS
Fruit (Continued)	
Pineapple	.26
Mango	.26
Apricots, canned in juice	.28
Grapes, Thompson, seedless	.32
Grains and Starchy Vegetables (Dry cereal = 1 oz cereal plus 1/2 cup nonfat milk)	
Oatmeal, one minute	.21
Cream of Wheat, Nabisco	.22
Barley, pearled	.24
Sweet potato, cooked	.27
Spaghetti, protein enriched	**.31**
Mueslix, Kellogg's, with milk	.32
Bran Chex, General Mills, with milk	.34
Raisin bran, with milk	.35
Yams, cooked	.35
Potatoes, French fried	.36
Buckwheat	.36
Mini Wheats, frosted, with milk	.37
Corn	.37
Fettucinne, egg	.37
Pizza, thin	**.37**
Spaghetti, whole wheat	.38
Cheerios, General Mills, with milk	.39
Special K, Kellogg's, with milk	.40
Potatoes, boiled (new potatoes are best)	.40
Protein Foods	
Peas, cooked	.21
Navy beans, cooked	.24
Yogurt, fruit on bottom	.25

LOW GLYCEMIC DENSITY FOOD (Continued)

Food	GDS
Protein Foods (Continued)	
Baked beans, canned	.30
Indulgences	
Ice cream, premium	**.28**
Pudding	.33
Ice cream, light, low fat	**.36**

MODERATE GLYCEMIC DENSITY FOOD

The moderate Glycemic Density list contains those foods with a Glycemic Density of greater than .40 but no more than .60. My evaluation is to exercise restraint with these foods.

Food	GDS
Fruits	
Banana	.41
Peaches, canned in heavy syrup	.42
Grains and Starchy Vegetables	
Potatoes, mashed with milk and butter	.45
Total, General Mills, with milk	.45
Shredded wheat, with milk	.45
Rice, brown	.47
Grapenuts, Post with milk	.47
Potatoes, mashed with milk	.48
Grape Nut Flakes, Post, with milk	.48
Spaghetti, white	.50
Golden Grahams, General Mills, with milk	.51
Coco Puffs, General Mills, with milk	.51
Corn flakes, with milk	.55
Indulgences	
Ice cream, regular	**.41**
Ice cream, nonfat, no sugar added	**.47**

HIGH GLYCEMIC DENSITY FOOD

The high Glycemic Density list contains those foods with a Glycemic Density greater than .60. My evaluation is to use these foods sparingly.

Food	GDS
Fruits	
Prunes	.61
Apricots, dried	.72
Cranberry sauce	.86
Dates	1.09
Figs, dried	1.39
Raisins	1.49
Grains and Starchy Vegetables	
Crispy rice, with milk	.61
Potatoes, Russet, baked	.63
Macaroni and Cheese, boxed	.65
Bread, whole grain wheat	.71
Rice, white	.76
Bread, pumpernickel	.77
Pancake, plain	.78
Bread, sourdough	.90
Waffle, plain	.96
Corn tortilla	1.02
Bread, seven grain	1.04
Bread, rye	1.08
Pita	1.14
Bread, wheat	1.25
Corn bread	1.27
Bread, white	1.35
Bagel	1.48
English muffin	1.49

HIGH GLYCEMIC DENSITY FOOD (Continued)

Food	GDS
Sugars Limit to 16 calories per serving: Note: Fructose is not the same as high fructose corn syrup. The latter is designed to emulate table sugar	
Fructose (fruit sugar)	.76
Jams and Jellies	1.43
Honey	2.20
Table syrup	2.44
Indulgences	
Oatmeal cookie, with raisins	.75
M & M Peanut candy	.78
Chocolate cake, with frosting	.85
Shortbread cookie	.85
Blueberry muffin	.85
Apple muffin	.97
Potato chips	1.00
Banana bread	1.00
Rye crisp bread	1.05
Power Bar	1.06
Pound cake	1.08
Twix cookie bar	1.18
Chocolate bar	1.26
Doughnut, cake	1.37
Corn chips	1.44
Soda crackers	1.48
Angel food cake	1.53
Snickers candy	1.56
Wheat thins	1.61
Whole grain bar, Kudos	1.66

HIGH GLYCEMIC DENSITY FOOD (Continued)

Food	**GDS**
Indulgences (Continued)	
Popcorn, air popped	1.80
Cupcake with frosting	1.83
Vanilla wafers	1.98
Graham crackers	2.12
Pretzels	2.61
Fruit roll ups	3.40

GLYCEMIC-DENSITY ENERGY-DENSITY COMPARISON

The energy density rankings are taken from the Volumetric series of books by Barbara Rolls of Pennsylvania State University. The individual values, however, may differ because I calculated the values from scratch. This I did out of copyright considerations.

As the Glycemic Density rises, many foods have a higher GDS ranking than their energy density counterpart. This is because these foods tend to have a higher Glycemic Index. For these it may be better to follow the Glycemic Density suggestions. ***See Part 3: Table 4 - Glycemic-Density Energy-Density Comparison*** for a more detailed analysis.

Portion Control Recommendations

GDS	ED	Recommendation
0 to .20	0 to .60	Very Low-Eat Satisfying Portions
.21 to .40	.61 to 1.50	Low-Begin Portion Control
.41 to .60	1.51 to 4.00	Moderate-Watch Portion Size
.61 +	4.00+	High-Limit Portions

Foods That Are More Restrictive With Energy Density

Food	ED	ED Rank
Very Low Glycemic Density		
Cottage cheese, nonfat	.71	Low
Cottage cheese, low fat (1%)	.72	Low
Kidney beans, canned	.84	Low
Pinto beans, canned	.86	Low
Cottage cheese, reduced fat (2%)	.88	Low
Cottage cheese, plain	1.03	Low
Custard	1.05	Low
Olives	1.14	Low
Lima beans, cooked	1.15	Low
Lentils, cooked	1.16	Low
Soybeans, fresh	1.47	Low
Avocados	1.67	Mod
Cashews (May also select other nuts)	5.72	High
Peanuts	5.83	High
Other nuts		High
Low Glycemic Density		
Spaghetti, protein enriched	1.64	Mod
Ice cream, light, low fat	1.77	Mod
Pizza, thin	2.48	Mod
Ice cream, premium	2.54	Mod

DESIGNING A DIET

The Glycemic Density food lists can be used in several types of diets. They compare quite favorably with the Carbohydrate Ladder used by those on a low carbohydrate diet. The participants on these diets begin eating foods on the lower rungs of the ladder and end up on a higher level according to their carbohydrate tolerance.

Carbohydrate Ladder (Used for Low Carb Dieting)

Rank	Food	GDS
12	Indulgences	.12 to 3.40
11	High carb fruits (canned fruit, dried fruit, bananas)	.32 to 1.49
10	Grains	.12 to 1.35
9	High carb vegetables (sweet potatoes, yams, white potatoes)	.27 to .63
8	Legumes (beans, peas, lentils)	.18 to .24
7	Moderate carb fruit (apples, oranges, peaches, grapefruit)	.10 to .20
6	Borderline vegetables (tomatoes, carrots, beets)	.00 to .14
5	Berries	.09 to .10
4	Milk and Yogurt	.05 to .06
3	Nuts and seeds	.00 to .06
2	Creams and hard cheeses	-
1	Low carbohydrate vegetables	-

Glycemic Density might be able to be used to design a more exact Carbohydrate Ladder, perhaps ascending from .00 to .10, .11 to .20, .21 to .30 and so forth. The condensed guide on the next page is broken down in this manner.

For traditional dieting, using Glycemic Density within the recommendations of the Mayo Healthy Eating Food Pyramid may be the ideal to follow. The minimum servings provide 1200 calories. Nutritionists recommend we eat at least this amount of calories. Here is a recap of these recommendations:

Mayo Clinic Minimum Food Recommendations

Food Group	Daily Servings	Serving Size
Indulgences	1	75 Calories a day or 525 a week.
Olive oil, nuts, canola oil, avocados.	3 - 5	45 Calories each such as 1 tsp oil or 2 tbsp nuts.
Meats, dairy and legumes.	3 - 7	110 Calories each such as 3 ounces of fish or poultry, 2 ounces of red meat, 1 cup of milk or yogurt, or 1/2 cup of beans.
Whole Grains and Starchy Vegetables	4 - 8	70 Calories each such as 1/2 cup of cereal, grains, vegetables or 1 slice of bread.
Fruits	3 or more	60 Calories each 1/2 cup fruit.
Low Carb (Non Starchy) Vegetables	4 or more	25 Calories each 2 cups raw leafy or 1 cup solid.

CONDENSED GUIDE

Low Carbohydrate Vegetables, GDS="-"			
Spinach Cauliflower Lettuce Radishes Cucumber	Mushrooms Celery Zucchini Brussels sprouts	Cabbage Tomatoes Asparagus Green pepper Green beans	Broccoli Snap Peas Green onion
Very Low Glycemic Density		**Low Glycemic Density**	
GDS 0 to .10	**GDS .11 to .20**	**GDS .21 to .30**	**GDS .31 to .40**
Fruits			
Strawberries Grapefruit	Cherries Peaches Peaches, in juice Pears, in juice Plumbs Oranges Apples Pears Cantaloupe Applesauce, natural	Watermelon Apricots Fruit cocktail, in juice Peaches, in lt syrup Kiwi Pineapple Mango Apricots, in juice	Grapes
Grains and Starchy Vegetables			
Squash Carrots	Zoom / Wheatena Beets Oatmeal, old fashioned All-Bran, with milk	Oatmeal, 1 min Crm of Wheat Barley, pearled Sweet potatoes Spaghetti, protein	Mueslix, with milk Raisin Bran, with milk Bran Chex, with milk Mini Wheats, with milk Cheerios, with milk Yams Potatoes, French fried Buckwheat Corn Fettucinne, egg Pizza Spaghetti, whole wheat Potatoes, boiled
Protein Foods			
Cottage cheese Milk Yogurt, no sugar	Most Beans	Peas Navy beans Yogurt, sugar & fruit Baked beans	
Fats	**Indulgences**		
Olives Avocados Nuts	Pudding, sugar free nonfat milk Custard	Ice Cream, premium	Pudding, regular Ice Cream light, low fat

CONDENSED GUIDE (Continued)

Moderate Glycemic Density		High Glycemic Density	
GDS .41 to .50	**GDS .51 to .60**	**GDS .61 to 1.00**	**GDS 1.01+**
		Fruits	
Bananas Peaches, in heavy syrup		Prunes, Apricots, dried Cranberry sauce	Dates Figs Raisins
		Grains and Starchy Vegetables	
Special K, with milk Total, with milk Shredded Wheat with milk Grapenuts, with milk Golden Grahams, with milk Potatoes, mashed Rice, brown Spaghetti, white	Coco Puffs, with milk Corn Flakes, with milk Couscous	Crispy Rice, with milk Potatoes, baked Macaroni and Cheese Bread, whole grain Rice, white Bread, pumpernickel Pancake, plain Bread, sourdough Waffle, plain	Corn tortilla Bread, seven grain Bread, rye Pita Bread, wheat Bagel English muffin
		Indulgences	
Ice cream, regular	Ice cream, nonfat, no sugar added	M&M Peanut Chocolate cake with frosting Shortbread cookie Blueberry muffin Apple muffin Potato chips	Pound cake Angel food cake Most cookies Crackers Donuts Most candy Cupcakes Corn chips Pretzels Pasties Pies
		Sugars	
		Fructose	Jams and Jellies Honey Table Sugar

PART II
DEVELOPMENT OF GLYCEMIC DENSITY

CHAPTER THREE: GLYCEMIC INDEX

DEVELOPMENT OF THE GLYCEMIC INDEX

The glycemic index (GI) was first developed by Dr. David Jenkins of the University of Toronto, Canada. The book The New Glucose Revolution – The Authoritative Guide to the Glycemic Index is the best resource for further study. It was written by Jennie Brand-Miller, Ph.D. of the University of Sydney, Australia, Thomas M.S. Wolever, M.D., Ph.D. of the University of Toronto, Canada, Kaye Foster-Powell, a practicing dietician in New South Wales, Australia and Stephen Colagiuri, M.D. of the Prince of Wales Hospital in Randwick, New South Wales. Authors who have contributed to other books in the series are Johanna Burani, a registered dietitian in Medham, New Jersey, Sussana Holt, Ph.D., head of the Glycemic Index Research Center of the University of Sydney and Philippa Sandall, a nutritionist author. There have also been substantial contributions by the Harvard Medical School under the direction of Walter C. Willett, M.D.

WHAT IS THE GLYCEMIC INDEX?

The glycemic index measures how rapidly carbohydrates are digested and contribute to a rise in blood sugar. Glucose is the standard reference and is given a value of one hundred. Other foods are given a ranking based on the percentage rise in blood sugar as compared to glucose. A glycemic index of forty, for example, means that a serving of a test food (containing twenty-five grams of carbohydrates) would cause a rise in blood sugar that is forty percent of that caused by a serving of glucose (containing twenty-five grams of carbohydrates). What happens to the rest of the carbohydrates? The body is continually using up energy. If a food takes longer to digest, some of the glucose in the blood will be drawn down by the body's energy requirements, and not as much will be left to elevate blood sugar levels. Some types of sugar, such as fructose in fruit and galactose in milk, are retained by the liver and are only released as the body requires.

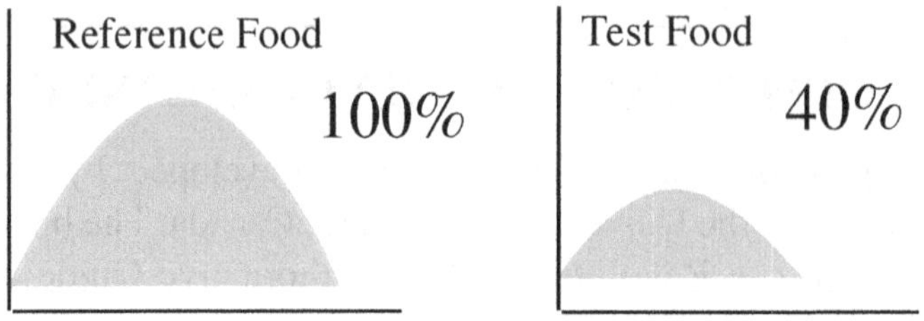

The Glycemic Index

THE GLYCEMIC LOAD

A concept related to the glycemic index is called the glycemic load (GL). This was developed by researchers at the Harvard School of Medicine. It is a measurement of the actual grams of carbohydrates that participate in the blood sugar rise for a given serving size. To calculate the glycemic load: the glycemic index is first converted to a decimal by dividing by one hundred; this is then multiplied by the grams of carbohydrates in the serving size. As an example, old fashioned oatmeal has a glycemic index of forty-two. A half cup serving of cooked oatmeal has twelve grams of carbohydrates. If we multiply the 12 grams by 42/100, we have a glycemic load of 5 grams. Of the twelve grams of carbohydrates in the oatmeal, only five grams will participate in a blood sugar rise.

Glycemic Load

Food	Serv Size	Net Carbs	GI	GL
Old Fashioned Oatmeal	1/2 cup	12	42	5

The importance of the glycemic load is that it allows us to eat foods with a higher glycemic index and still monitor their impact upon our blood sugar. High glycemic foods do not need to be avoided altogether. Many contain high levels of essential vitamins and minerals and can contribute to overall health. Low glycemic foods may also have higher levels of calories, due to their protein and fat content.

We can also use the glycemic load to determine the total glycemic index of a meal. The formula for calculating the glycemic index of a meal is:

Glycemic-Index = Total-Glycemic-Load / Total-Net-Carbohydrate-Grams.

Net carbohydrates are the total carbohydrates less the fiber. After this calculation, we then convert the decimal to a percentage. This is done by multiplying the result by 100.

The Glycemic Index of a Meal

Food	Serv Size	Net Carbs	GI	GL
Orange juice	1/2 cup	13	46	6
Corn Flakes	1 oz	24	84	20
Milk	1/2 cup	6	27	2
1 slice of toast	1 oz	13	70	9
Total for meal		**56**	**66**	**37**

The total glycemic load for this meal was 37. If we divide this number by the total net carbohydrates of 56, we get the decimal figure .66. Multiplying this by 100 yields the glycemic index of 66 for the meal.

Glycemic-Index of 66 =
(Total-Glycemic-Load of 37
*/ Total-Net Carbohydrate-Grams of 56) * 100.*

There are three drawbacks to this method.

(1) Cooking tends to alter the glycemic index. Usually, cooking will quicken the digestive process and will lead to a raising of the glycemic index. This is why it is always best to eat foods in their natural state whenever possible.

(2) Even though combining carbohydrates can lower the overall glycemic index, the glycemic load is always accumulative.

(3) The addition of acids, fats and proteins also alters the glycemic index. Thus, if we add a serving of meat or an acid such as lemon juice or vinegar to the meal, we will lower the overall glycemic index as well as the corresponding glycemic load. No formula for calculating this change has been developed. The only sure way is to test the effect on live subjects by measuring their blood sugar levels at varying intervals, as has been done with pizza, for example.

WHAT CONSTITUTES A HEALTHY GLYCEMIC INDEX?

Experts have established the following glycemic index guidelines. A glycemic index of 55 or under is considered low. A glycemic index of 56-69 is considered moderate. Any glycemic index of 70 or greater is considered high. Glycemic load guidelines have also been established. A glycemic load of 10 or less for a particular serving is considered low. A glycemic load of 11-19 is considered moderate. Any glycemic load of 20 or greater is considered high.

Another important glycemic index is that of 40. If we eat a food with a GI less than 40, we can also eat a high GI food with a similar amount of carbohydrates, and still have an overall low glycemic index.

Total Glycemic Index = 40 + 70 = 110

Average Glycemic Index = 110 / 2 = 55

Glycemic Index and Glycemic Load Guidelines

Glycemic Index		Glycemic Load	
Low	0-55	Low	0-10
Moderate	56-69	Moderate	11-19
High	70+	High	20+

HOW DOES THE GLYCEMIC INDEX HELP US?

We should select the majority of our carbohydrate foods from those which have a low glycemic index. When we choose to eat foods that have a moderate or high glycemic index, we should apportion our serving sizes, so that the overall glycemic load remains low. Total calories are also a consideration. Low glycemic foods often have a high calorie level which can contribute to weight gain. This is a topic that will be discussed in our next chapter.

By eating foods that have a low glycemic impact, we are better able to control our blood sugar level. With blood sugar levels under control, we do not get hungry as often and will not eat as much. The glycemic index does yield some surprises from what was previously thought. For example, it was thought that complex carbohydrates, such as potatoes, would take longer to digest than simple sugars, such as sucrose or table sugar. However, under actual testing, this premise has been shown to be false. This is because common table sugar consists of a unit of glucose and a unit of fructose. While glucose has a glycemic index of one hundred, fructose is quite low on the glycemic index at nineteen, giving an overall average of sixty. A baked potato contains a long string of glucose units which are rapidly broken down at a glycemic index of eighty-five.

One of the most important factors in the glycemic index is the amount and type of fiber. If the nutrient part of the food is physically entrapped within a fibrous coating, it will take longer to digest. How rapidly the starch particle absorbs water is another factor. Larger particles absorb water less rapidly and take longer to digest. The less refined the food, the longer the digestion. Some other factors, which cause the glycemic index to vary, are the amount of protein, the amount of fat and the acidity. It takes longer to digest protein and fat than it does carbohydrates. Acid, such as is found in oranges or lemons, also lengthens the digestive process as they have to be neutralized.

Factors that Determine the Glycemic Index

Starch gelatinization
The less swollen the starch, the slower the rate of digestion.
Physical Entrapment
Fiber surrounding the nutrient slows the rate of digestion.
Type of Fiber
Amylose fibers, which have straight chains of molecules, digest more slowly than amylopectine, which have branching chains.
Particle Size
The larger the food particle, the slower the rate of digestion.
Amount of Sugar
Table sugar (sucrose) consists of 50% glucose and 50% fructose, whereas starch consists of 100% glucose. The presence of sugar slows down the digestion rate of starches.
Fructose to Sucrose Ratio in Fruit
Sucrose has a glycemic index of 61, whereas fructose has a glycemic index of only 19. The greater the proportion of fructose, the lower the glycemic index of the fruit.
Acidity
Acids such as vinegar, lemon juice, lime juice and brine have to be neutralized which slows down digestion.
Fat and Protein
Fat and protein slows down the rate of stomach emptying which in turn slows down digestion.

GLYCEMIC INDEX OF SOME POPULAR FOODS

The following is a brief list of the glycemic index and load of some popular foods.

Glycemic Index and Load of Popular Foods

Food	Serv Size	Net Carb	GI	GL
Baked Goods				
Angel food Cake	1/12 cake	31	67	21
Choc cake with frosting	1/12 cake	32	38	12
Beverages				
Grape juice	1/2 cup	16	60	10
Orange juice	1/2 cup	14	53	7
Tomato juice	1/2 cup	5	33	2
Breakfast Cereals				
All Bran, Kellogg's	1/2 cup	14	38	5
Corn flakes	1/2 cup	18	81	15
Oatmeal, old fashioned	1/2 cup	12	42	5
Oatmeal, one minute	1/2 cup	12	66	8
Shredded Wheat	1/2 cup	18	75	14
Other breakfast food				
Pancakes	1 pancake	13	67	9
Waffles	1 square	13	76	10
Cookies and crackers				
Graham crackers	4 squares	20	74	15
Oatmeal cookie	1 oz	21	55	12
Soda crackers	1 oz	18	74	13
Vanilla wafers	1 oz	24	77	17

Glycemic Index and Load of Popular Foods (continued)

Food	Serv Size	Net Carb	GI	GL
Dairy				
Milk, skim	1 cup	12	30	4
Milk, whole	1 cup	11	27	3
Chocolate milk (1%)	1 cup	26	34	9
Ice cream, regular	1/2 cup	17	38	6
Ice cream, fat free, no sugar added	1/2 cup	18	50	10
Yogurt, fruit flavor, low fat, no sugar added	3/4 cup	16	14	2
Yogurt, low fat, fruit on bottom	3/4 cup	32	31	10
Fruit				
Apples	1/2 cup	8	38	3
Bananas	1/2 cup	26	50	12
Oranges	1/2 cup	6	42	3
Raisins	1/4 cup	28	64	18
Strawberries, fresh	1/2 cup	4	40	2
Grains				
Corn	1/2 cup	12	54	7
Rice, brown	1/2 cup	20	44	11
Rice, white	1/2 cup	26	69	18
Legumes				
Baked beans, canned	1/2 cup	16	48	8
Peas, fresh, cooked	3/4 cup	7	42	3
Soy beans, boiled	1/2 cup	6	18	1
Pasta				
Spaghetti, white	1/2 cup	19	44	8
Spaghetti, whole wheat	1/2 cup	17	37	6

Glycemic Index and Load of Popular Foods (continued)

Food	Serv Size	Net Carb	GI	GL
Pizza				
Pizza Hut Super Supreme, pan	1 med slice	27	36	10
Pizza Hut Super Supreme, thin	1 med slice	21	39	8
Snacks				
Chocolate bar	1 oz	15	43	6
Corn chips	1 oz	16	63	10
Jelly beans	15 items	15	78	12
Kudos whole grain bar	1 bar	17	62	11
M & M Peanut	1 oz	18	33	6
Peanuts	1 oz	6	14	2
Popcorn, plain, popped	2 cups	8	72	6
Potato chips	1 oz	14	54	8
Soups				
Chicken noodle soup	1 cup	10	1	-
Minestrone soup	1 cup	10	39	5
Split pea soup	1 cup	26	60	16
Tomato soup (with water)	1 cup	16	38	6
Tomato soup (with nonfat milk)	1 cup	28	36	10
Vegetables				
Carrots, raw	1/2 cup	4	47	2
Potatoes, baked Russet	1/2 cup	19	85	16
Spinach	1 cup raw	2	0	0
Sweet Potato	1/2 cup	20	44	9
Tomatoes	1 cup raw	6	16	1
Zucchini	1 cup raw	2	-	-

CHAPTER FOUR: ENERGY DENSITY

WHAT IS ENERGY DENSITY?

One of the major methods of managing weight is to eat foods with a lower energy density. By eating foods with fewer calories per serving, we are able to eat our fill without feeling deprived. For example, all of the following food servings have 110 calories.

Energy Density Comparison of Foods with 110 Calories

2 Fig Newtons
1/2 cup of pork and beans
1 cup of Cheerios
1 3/4 cups of chopped apples
2 1/4 cups of strawberries
3 2/3 cups of popcorn
5 1/2 cups of chopped celery
9 cups of spinach
11 cups of shredded lettuce

We could easily eat a half dozen fig newtons without thinking, but can anybody sit down and eat eleven cups of shredded lettuce? The more foods that we can eat with a lower energy density, the easier it will be for us to manage our weight.

GLYCEMIC LOAD AND ENERGY DENSITY

When we add the factor of energy density to that of the glycemic index, it alters whether or not a food is really a healthy choice. A piece of angel food cake has a glycemic load of twenty-one while a piece of chocolate cake has a glycemic load of only thirteen. From a blood sugar point of view, the chocolate cake would be the desirable thing to eat.

But the reason that the chocolate cake has a lower glycemic index, is because the presence of fat lengthens the digestion process. From an energy density point of view, the angel food cake has a calorie count of only 73, while the piece of chocolate cake has a whopping 264 calories! The situation is even worse when it comes to pizza. A piece of Pizza Hut Super Supreme thin pizza has a glycemic load of only eight, but the calories add up to 280.

Energy Density and Glycemic Load Comparison

Food	Cal	Serv Size	Net Carbs	GI	GL
Angel food cake	73	1/12 cake	31	67	21
Chocolate cake with frosting	264	1/12 cake	33	38	13
Pizza Hut Super Supreme, thin	280	1 med slice	21	39	8

In the short run, especially for diabetics, it may be better to take the low glycemic choice. But since weight is such a large factor in diabetes, the long term choice of reducing overall calories must also be a consideration. A truly successful weight management plan must aim at eating foods that are both low glycemic and low calorie.

THE ENERGY DENSITY OF FOOD GROUPS

The Mayo Clinic has developed a food pyramid which ranks foods according to energy density. The food group with the lowest energy density are those known as non-starchy or low carbohydrate vegetables such as broccoli, cauliflower, asparagus and salad vegetables. A typical serving would have about twenty-five calories per cup. This group is placed at the bottom of the pyramid. Above the vegetables are the fruits which contain about sixty calories per half cup. Denser still are the grain products and starchy vegetables, such as potatoes and yams, which have about seventy calories per half cup or one ounce serving. Protein sources and fats are even denser.

We need to eat foods from each of the levels in order to supply us with all of the nutrients that our bodies require. After that, if we are still hungry, we should select foods from the lower levels of the pyramid.

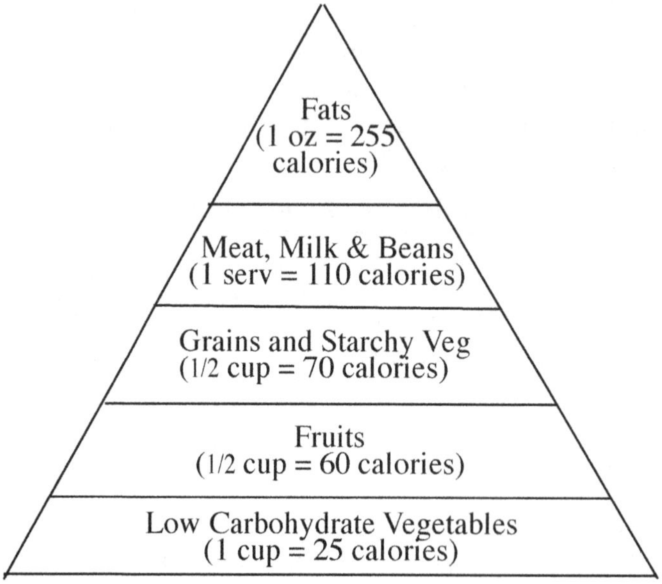

Energy Density Analysis of Food Groups

THE TARGET CONCEPT

Energy density amounts can very significantly within each of the categories. Processed grain products may have added sugar and fat. Meat and milk products have a fat content that varies over a wide range. A good analysis of energy density is the book Fit or Fat Target Diet by Covert Bailey, an expert in the field of fitness, and its successor Smart Eating coauthored with nutritionist Ronda Gates. In them, he took the pie chart containing the four food groups, which later developed into the food pyramid, and expanded it into a target. In the bull's eye were those foods in each group that were lowest in calories, and in the case of carbohydrates, high in fiber. Each subsequent ring contains foods in each group that were higher in calories and/or lower in fiber.

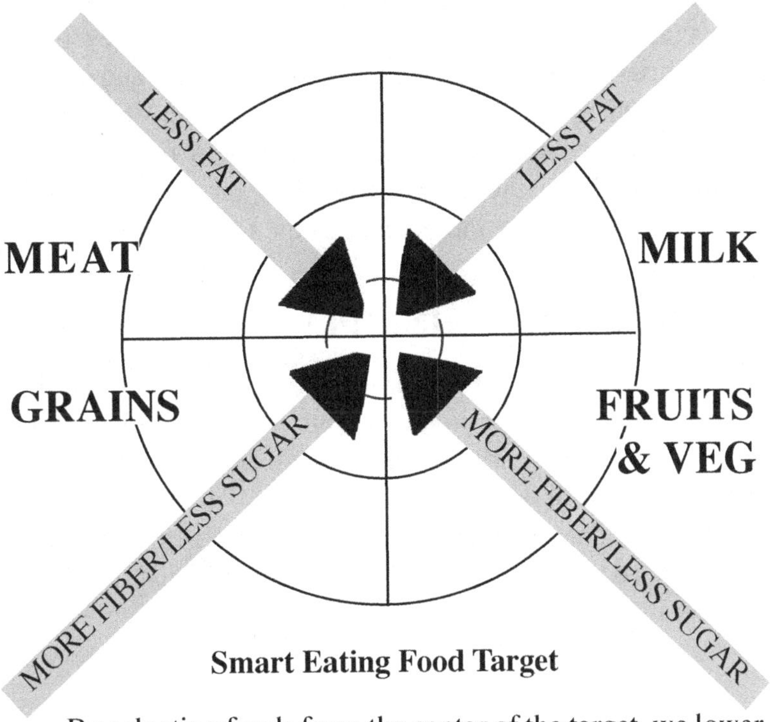

Smart Eating Food Target

By selecting foods from the center of the target, we lower both the energy density and the glycemic load of the foods which we choose to eat.

MEASURING ENERGY DENSITY

Much of the credit for furthering the research on energy density goes to Barbara Rolls of Pennsylvania State University. She is the author of the Volumetrics Weight-Control Plan and the Volumetrics Eating Plan, both published by Harper-Collins. Through research it was discovered that most people tend to eat the same weight of food each day. By lowering the energy density of our diet, it is possible to be satisfied with fewer calories. This will help us to both lose weight, and once we have reached our weight goal, to keep our weight at a healthy level.

It was discovered that an actual number could be used to classify the energy density of foods. The formula is to divide the calories of the serving size by the weight in grams. The result is a number from zero to nine. An energy density of zero means that the food contains no calories. An energy density of nine means that the food is pure fat. Carbohydrates and protein each contain four calories of energy, hence pure protein or carbohydrate would have an energy density of four. The less fat that a food has, the lower will be its energy density. Also the more fiber and the more water in the food, the lower will be the energy density.

APPLYING THE PRINCIPLE

The single most beneficial thing I did was to divide my plate into three sections. On half the plate I placed one to two cups of low carbohydrate vegetables. A quarter of the plate contained about three fourths cup of grains or starchy vegetables. The remaining quarter contained a protein selection. To this I added an in-between meal fruit snack and a moderate evening dessert.

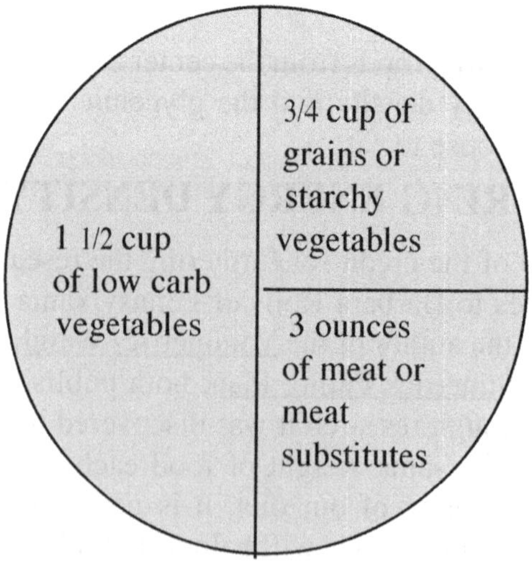

Ideal Portions of Food for a Meal

I have found this idea mentioned in at least three books, The GI Diet by Rick Gallup, The Simple 0-to-10 GI Diet by Azmina Govindji and Nina Puddefoot, and The Holford Low GL Diet by Patrick Holford.

The serving sizes suggested fulfil the minimum nutritional recommendations of the Mayo Clinic. These are four or more servings of low carbohydrate vegetables, three or more servings of fruit, four servings of grains or starchy vegetables, three servings of protein, three servings of healthy fat and no more than 75 calories a day for an indulgent dessert.

LOSING WEIGHT BY EATING MORE

By using the concept of energy density as a guide to managing our weight, we are able to reverse the usual starvation method of dieting. We can actually lose weight by eating more of the right kinds of food. By eating low energy density foods, we become satisfied with fewer calories.

Of course we could carry this to an extreme. We might reason that we could lose weight even more rapidly by eating nothing but low carbohydrate vegetables. But if we did so, our health would definitely suffer. We might have plenty of vitamins and minerals, but we would lack sufficient carbohydrates to power our brains. We would also lack sufficient protein to replace lost muscle mass, including the muscle mass of our heart and other vital organs. We could probably give our brains adequate nourishment by eating fruits as well as low carbohydrate vegetables, but we would still suffer from lack of protein. Most quality protein sources are high energy density. Only by combining good nutrition with calorie and carbohydrate management can we be successful.

CHAPTER FIVE: GLYCEMIC DENSITY

BRIEF OVERVIEW

What I call Glycemic Density is a listing of foods that provide the most hunger satisfaction with the least glycemic impact. The way this is measured is to calculate the carbohydrate calories per gram of food and multiply the result by the glycemic index expressed as a decimal.

GLYCEMIC INDEX

The Glycemic Index (GI) is a measurement of how rapidly various carbohydrate foods are digested and contribute to a rise in blood sugar. Foods are ranked on a scale from 0 to 100. Glucose is assigned a value of one hundred because it is the food with the greatest impact on blood sugar. A related concept is the Glycemic Load (GL). It apples the glycemic index to a particular serving size. The glycemic load is calculated by converting the glycemic index to a decimal and multiplying by the carbohydrate grams in the serving. In other words:

Glycemic-Load = Carbohydrate-Grams
*** Glycemic-Index / 100.**

See chapter three for a more detailed description of the glycemic index and the glycemic load.

ENERGY DENSITY

Energy or calorie density is not necessary in order to calculate the Glycemic Density, but it does have a logical relationship. Energy density is a calculation of calories per gram of food. It is calculated by dividing the total grams of the serving size by the total calories. The formula is:

Energy-Density = Calories / Total-Grams-in Serving

See chapter four for a more detailed description of energy density. As mentioned earlier, research has revealed that people tend to eat the same weight volume of food regardless of the calorie content. By choosing foods with a lower energy density, weight management becomes much simpler. Glycemic Density also takes advantage of this principle.

Glycemic Density combines the measurement of carbohydrate calories per gram of food with the glycemic index of the particular food. There are two factors which will cause a change in the Glycemic Density of a food. One of these is the glycemic index. The comparison of several sugars is a good way to illustrate this. Pure glucose would have a Glycemic Density of four, because a gram of glucose would have four calories with a glycemic index of 100%. A gram of regular table sugar also has four calories, but it only has a glycemic index of 61%. Consequently, it only has a Glycemic Density of 2.44. A gram of fructose also has four calories. It's glycemic index is 19% which gives it a Glycemic Density of .76.

(1) Comparison of Food with Different Glycemic Indexes
(GI=Glycemic Index, GDS=Glycemic Density Score)

Food	Calories	Total Grams	Carb Grams	GI	GDS
Glucose	4	1.0	1.0	100	4.00
Sucrose (table sugar)	4	1.0	1.0	61	2.44
Fructose (fruit sugar)	4	1.0	1.0	19	.76

The other way is by changing the weight of the food. This can happen with added fat, protein, fiber or water. Fat and protein of course will add additional calories. The addition of fiber and water have the advantage of increasing the weight of the food without adding calories. Here is a comparison of several foods with the same glycemic index but different weights.

(2) Comparison of Food with Different Weights
(GI=Glycemic Index, GDS=Glycemic Density Score)

Food	Calories	Total Grams	Carb Grams	GI	GDS
Twix cookie	7.5	1.5	1.0	44	1.18
Spaghetti, white	5.5	3.5	1.0	44	.50
Pudding, sugar free, nonfat milk	7.9	14.3	1.0	44	.12

Notice that the serving of pudding provides about the same amount of calories as the Twix cookie, but the weight of the pudding is almost ten times as much. That means that the pudding will provide ten times the hunger satisfaction with the same level of calories and a far lower glycemic impact.

WHAT ARE THE ADVANTAGES OF GLYCEMIC DENSITY?

Much of the research into energy or calorie density was conducted by Barbara Rolls of Pennsylvania State University. Her research indicated that people tend to eat the same weight of food over a two or three day period. That being true, if we select the majority of our food from those that have a lower Glycemic Density, we would be able to lower the overall glycemic impact without having to worry about measuring the specific glycemic index or glycemic load. This would make our food choices much easier, especially for diabetics and others who may be carbohydrate sensitive. We would simply have to choose foods that are lower on the Glycemic Density Score.

Many of these foods are vegetables and fruits. As well as being low Glycemic Density, these foods are filled with vitamins and minerals. They are also rich in what are called phytochemicals. This has a lot to do with the color of foods.

Phytochemicals provide powerful cardiovascular and anti-cancer properties. See ***Part 4: Table 8 - Super Foods*** for a list of these foods.

EXTRACTING THE FOOD CALORIE AND WEIGHT VALUES

The first thing that I did in calculating the Glycemic Density values was to make a spread sheet containing the food name, serving size, calories, total weight in grams and net carbohydrates in grams (subtracting the fiber). I acquired most of this information from the National Nutrient Data Base which can be found at www.nal.usda.gov/fnic/foodcomp/search. If it was not available there, I acquired the information from actual food labels. For example, below is the information extracted for old fashioned oatmeal.

CAL=Calories, GM=Total weight in grams
CARB=Net Carbohydrate Grams.

FOOD	SIZE	CAL	GM	CARB
Oatmeal, old fashioned	1 cup	150	280	23

See ***Part 3: Table 1 - Extraction of information*** for a complete list. The foods selected are somewhat restricted because not every food has been tested to determine its glycemic index. Glycemic index is a necessary part of the Glycemic Density formula.

CARBOHYDRATE DENSITY

The next thing is to calculate the carbohydrate density. This is the measurement of carbohydrate calories per gram of food. Carbohydrate calories can be calculated by first multiplying the net carbohydrate grams (without the fiber) by four, because there are four calories per gram of carbohydrate. This is then divided by the total grams in the serving. This yields a number between zero and four. Pure sugar would have a carbohydrate density of four. As with the energy density, the presence of fiber and water lowers the carbohydrate density. The presence of fat and protein also lowers the carbohydrate density. The complete formula is:

Carbohydrate-Density = (Net-Carbohydrate-Grams *4) / Total-Grams-in-Serving.

Going back to our example of oatmeal, the bowl of oatmeal has 23 grams of carbohydrate and weighs 280 grams. The carbohydrate density would be 23 * 4 / 280 or about .33.

GLYCEMIC DENSITY SCORE

Carrying this one step further, if we multiply the carbohydrate density by the glycemic index expressed as a decimal, we have a glycemic load type number independent of the serving size, which I call the Glycemic Density Score (GDS). This gives a measurement of the glycemic impact as compared to the weight of the food. This measurement also ranges between zero and four. Four would be the value of pure glucose. The formula is:

GDS = Carbohydrate-Density * Glycemic Index / 100.

The complete formula is:

GDS = (Net-Carbohydrate-Grams * 4
/ Total-Grams-in-Serving)
*** (Glycemic-Index / 100).**

For our oatmeal, the Glycemic Density Score would be (23 * 4 / 280) * (42 / 100) or about .14.

CD=Calculated Carbohydrate Density. GI=Glycemic Index, GDS=Glycemic Density Score

FOOD	GM	CARB	CD	GI	GDS
Oatmeal, old fashioned	280	23	.33	42	.139

See ***Part 3: Table 2 - Calculation of Carbohydrate Density and Glycemic Density Score*** for a complete list. This particular table is sorted by carbohydrate density (CD). There are many people who are interested in carbohydrate management without the consideration of the glycemic index. This table is for them.

CREATING THE 100 GRAM SERVING LIST

The next thing was to convert all of the entries in the table to a weight measurement of 100 grams. By doing this it can be observed that the glycemic load parallels the Glycemic Density Score. This can be found in ***Part 3: Table 3 - 100 Gram and Cereal Lists.***

Food	**Cal**	**Net Carb**	**GI**	**GL**	**GDS**
Oatmeal, old fashioned	54	8.3	42	3	.14

ANALYSIS OF BREAKFAST CEREALS

Many dry breakfast cereals score rather poorly. However, most cereals are so light that even a small amount of milk dramatically improves the Glycemic Density Score. The same thing would be true when foods are combined with added water into a soup, or a fruit salad is made with sugar free gelatin. Sodas and fruit drinks generally satisfy thirst rather than hunger and are not listed in the tables. It is best to use them sparingly. The breakfast cereal comparison is also found in Table 3. Here is the listing for an ounce of Kellogg's All Bran without milk.

Food	**Cal**	**Net Carb**	**GI**	**GL**	**GDS**
All Bran, Kellogg's (28.35 grams)	74	12.0	38	4	.64

And here is the listing with 1/2 cup of nonfat milk.

Food	**Cal**	**Net Carb**	**GI**	**GL**	**GDS**
All Bran, Kellogg's (152 grams)	119	18.0	33	6	.16

The addition of the half cup of milk greatly improved its Glycemic Density Score.

GLYCEMIC DENSITY VERSES ENERGY DENSITY

Energy Density and Glycemic Density represent different philosophies of eating. The best way to reduce overall calories is through a low energy density diet. This allows a person to eat the maximum volume of food with a minimum of calories. The approach of low glycemic diets such as a low glycemic index diet, a low glycemic load diet or a low Glycemic Density diet is to effect weight management through carbohydrate control. Glycemic Density, in particular, allows a person to eat the maximum volume of food with the minimum glycemic impact. Yet, the Glycemic Density approach does compare favorably with the low energy density approach with a few exceptions, giving us the best of two worlds. See ***Part 3: Table 4 - Glycemic-Density Energy-Density Comparison.***

Food	ED #	ED Level	Reason
Very Low Glycemic Density			
Kidney beans, canned	.84	Low	Protein

The kidney beans in this example scores at a higher level with energy density than it does with Glycemic Density. This is because energy density counts the calories in fat and protein while Glycemic Density does not. A person who chooses to diet with calorie management rather than carbohydrate management might elect to use the energy density ranking.

Food	ED #	ED Level	Reason
Low Glycemic Density			
Watermelon	.30	Very Low	Glycemic

Watermelon, on the other hands scores at a higher level with Glycemic Density than it does with energy density. This is because watermelon has a very high glycemic index. A person may be better off using the Glycemic Density approach in this case.

Part 3 also contains two other useful tables. ***Part 3: Table 5 - List of Foods in Food Exchange Calories*** gives a practical list for those who are using the food exchange listings for calorie or carbohydrate management. The foods are broken down into Low Carbohydrate Vegetables (25 calories per serving), Fruits (60 calories), Grains and Starchy Vegetables (80 calories), Protein selections (110 calories), Sugars (16 calories), Fats (45 calories) and indulgences (100 calories). The foods are ranked in each group by the Glycemic Density Score.

Fruits (60 Calories)

FOOD	SIZE	MEAS	GM	CARB	GDS
Very Low Glycemic Density					
Strawberries	1.30	Cup	187	10.4	.09

Part 3: Table 6 - Glycemic Density Alphabetical List contains foods listed in alphabetical order showing the food name and the Glycemic Density Score.

HOW TO ADD GLYCEMIC DENSITIES

In order to add the Glycemic Densities of two foods, the following steps need to be followed:

(1) The weight (in grams) of the two foods needs to be added together.

(2) The carbohydrate content of the two foods (in grams) needs to be added together.

(3) The glycemic load of the two foods needs to be added together. The glycemic load is calculated by multiplying the net carbohydrate content of the food by the glycemic index expressed as a decimal.

(4) A new glycemic index needs to be calculated by dividing the glycemic load by the carbohydrate content. This is then multiplied by 100 to create a percentage.

(5) The Glycemic Density Score can now be calculated by applying the standard formula.

GDS = (Net-Carbohydrate-Grams * 4
/ Total-Grams-in-Serving)
*** (Glycemic-Index / 100).**

Here is an example of adding a cup of milk to a cup of old fashioned oatmeal which has been prepared with water.

Adding Glycemic Densities

Food	Serv	GM	Carbs	GI	GL	GDS
Oatmeal, Old fashioned	1 cup	280	23	42	9.66	.139
Milk, 1%	1 cup	244	12	29	3.48	.058
Combined		524	35	36	13.14	.096

Notice that the combined GDS is less than the old fashioned oatmeal by itself. This is generally the case when a lower density food is added to a higher density food.

This is not the case, however, when one food is substituted for another. Below is an example of old fashioned oatmeal made with milk rather than water. This is known in some countries as porridge.

Substituting Glycemic Densities

Food	Serv	GM	Carbs	GI	GL	GDS
Oatmeal, Old fashioned	1/2 cup dry	40	23	42	9.66	.139
Milk, 1%	1 cup	244	12	29	3.48	.058
Combined		284	35	36	13.14	.177

In this case where a lower density nutrient (water) is replaced by a higher density food (milk), the result is a food with a higher Glycemic Density.

PROTEIN FOODS

As mentioned earlier, most protein foods have a fairly low Glycemic Density Score. The above tables only consider those with at least some carbohydrates. It does not include meat, fish or poultry. Yet these foods are part of most everyday diets. It is generally recommended that we choose protein selections with as little fat as possible. Our table approaches protein from the opposite direction by emphasizing foods with a maximum percentage of calories from protein. This screens out foods with higher amounts of carbohydrate as well as fat. ***Part 4: Table 7 - Protein Density Foods*** contains a complete list of these foods. There is also a table designed specifically for vegetarians helping them make better protein selections.

MOVING FORWARD

Since research has concluded that people tend to eat the same weight volume, eating foods with a lower Glycemic Density can dramatically lower the glycemic impact of those foods which are eaten. This may be especially beneficial to diabetics and others who are carbohydrate sensitive. It may be easier to refer to a food list than worrying about calculating specific glycemic indices and specific glycemic loads.

The Glycemic Density Score came about through my own research, However, I am not a nutritional professional nor do I have the resources of a major university or health clinic. My hope is that those who have these resources will be willing to take up this principle and thoroughly evaluate it.

One of the things that need to be evaluated are food categories. What is a low Glycemic Density Score?; what is a moderate score?; and what is a high score? It appears to me that some good divisions would be: Very Low = 0 to .20, Low = .21 to .40, Moderate = .41 to .60, and anything over .60 would be high.

Other questions need to be answered. How much food should a person be able to eat from each category? Should there be some sort of diet plan? These are the sort of things that need to be determined.

PART III
GLYCEMIC DENSITY TABLES

TABLE ONE: EXTRACTION OF INFORMATION

CAL = Total Calories, GM = Total Grams,
CARB = Carbohydrate Grams without fiber,
Note: Dry cereal listings are without milk. Table is in Glycemic Density order (See Table 3: 100 Gram List).

FOOD	SIZE	CAL	GM	CARB
Spinach, cooked	1 cup	41	180	2
Cauliflower, cooked	1 cup	28	124	2
Lettuce, iceberg, raw	1 cup	10	72	1
Mushrooms, cooked	1 cup	28	108	2
Radishes, raw	1 cup	19	116	2
Cucumber, raw, sliced	1 cup	14	119	2
Celery, raw, chopped	1 cup	16	101	2
Lettuce, Romaine, raw	1 cup	8	47	1
Zucchini, cooked	1 cup	28	180	4
Brussels sprouts, cooked	1 cup	56	156	4
Cottage Cheese, plain	1/2 cup	116	113	3
Cottage Cheese, reduced fat (2%)	1/2 cup	101	113	3
Cottage Cheese, low fat (1%)	1/2 cup	82	113	3
Cabbage, raw, shredded	1 cup	18	70	2
Tomatoes, raw, chopped	1 cup	32	180	5
Cauliflower, raw	1 cup	25	100	3
Asparagus, cooked	1 cup	40	180	6

FOOD	SIZE	CAL	GM	CARB
Green pepper, raw, chopped	1 cup	30	149	5
Tomatoes, cooked	1 cup	43	240	8
Spinach, raw	1 cup	7	30	1
Broccoli, cooked	1 cup	54	156	6
Green beans, cooked	1 cup	38	135	5
Olives	10 large	50	44	2
Avocados	1/2 avocado	114	68	3
Broccoli, raw	1 cup	31	91	4
Cottage Cheese, nonfat	1/2 cup	80	113	5
Snap peas, raw, whole	1 cup	26	63	3
Green onions, raw	1 cup	32	100	5
Squash, winter, cooked	1/2 cup	56	102	10
Peanuts	30 nuts	166	28	6
Milk, soy	1 cup	112	262	10
Milk, whole	1 cup	146	244	11
Milk, reduced fat (2%)	1 cup	122	244	12
Soybeans, fresh	1/2 cup	188	128	9
Carrots, raw, chopped	1 cup	52	108	3
Milk, lowfat (1%)	1 cup	102	244	12
Cashews	18 nuts	163	28	13
Milk, nonfat	1 cup	91	245	13
Milk, chocolate, nonfat, no sugar added	1 cup	130	250	18
Yogurt, nonfat flavored, no sugar added	1 cup	100	226	13
Yogurt, nonfat, plain	1 cup	127	227	17
Kidney beans, canned	1/2 cup	108	128	10
Strawberries	1 cup	46	144	8

FOOD	SIZE	CAL	GM	CARB
Grapefruit	1 cup	74	230	16
Zoom or Wheatena	2/3 cup	140	195	23
Lentils, cooked	1/2 cup	115	99	12
Cherries, no pits	1 cup	91	145	20
Pudding, sugar free nonfat milk	1/2 cup	70	128	9
Milk, chocolate (1%)	1 cup	158	250	26
Beets, canned, drained	1 cup	53	170	9
Peaches	1 cup	60	154	13
Peaches, in juice	1/2 cup	50	124	12
Plumbs	1 cup	76	165	17
Pears, in juice	1/2 cup	60	125	12
Apples	1 cup	57	110	12
Oatmeal, old fashioned	1 cup	150	280	23
Oranges	1 cup	81	165	17
Lima beans, cooked	1/2 cup	108	94	13
Pinto beans, canned	1/2 cup	103	120	13
Custard	1/2 cup	148	141	15
Pears	1 cup	81	140	17
Cantaloupe	1 cup	54	160	12
Applesauce, no added sugar	1/2 cup	70	126	16
Apricots	1 cup	79	165	15
Peas, cooked	1/2 cup	116	98	12
Watermelon	1 cup	46	154	11
Oatmeal, one minute	1 cup	140	279	22
Cream of Wheat, Nabisco	1 cup	120	273	23
Fruit cocktail, in juice	1/2 cup	60	124	13
Peaches, in light syrup	1/2 cup	60	124	14

FOOD	SIZE	CAL	GM	CARB
Navy beans, cooked	1/2 cup	128	91	14
Barley, pearled	1/2 cup	96	78	19
Yogurt, fruit on bottom	1 cup	230	245	46
Kiwi	1 cup	108	177	21
Mango	1 cup	107	165	25
Pineapple	1 cup	74	155	17
Sweet potato, cooked	1 medium	115	151	23
Apricots, in juice	1/2 cup	60	124	14
Ice cream, premium	1/2 cup	188	74	14
Baked beans, canned	1/2 cup	120	127	20
Spaghetti, protein	1/2 cup	115	70	21
Grapes, Thompson, seedless	1 cup	104	151	26
Pudding, regular	1/2 cup	135	145	27
Yams, cooked	1/2 cup	79	68	16
Potatoes, instant milk/butter	1/2 cup	102	105	11
Ice cream, light, low fat	1/2 cup	110	62	15
Potatoes, French fried	10 fries	134	100	12
Buckwheat	1/2 cup	78	84	14
Corn	1/2 cup	66	77	13
Pizza, thin, 14 "	1 piece	305	123	29
Fettucinne, egg	1/2 cup	110	90	21
Spaghetti, whole wheat	1/2 cup	87	70	18
Potatoes, boiled	1/2 cup	68	78	14
Banana	7 to 8 inch	105	118	24
Ice cream, regular	1/2 cup	143	66	18
Peaches, in heavy syrup	1/2 cup	100	127	23

FOOD	SIZE	CAL	GM	CARB
Potatoes, mashed milk / butter	1/2 cup	119	105	16
Ice cream, nonfat, no sugar added	1/2 cup	92	68	16
Rice, brown	1/2 cup	109	98	21
Potatoes, mashed with milk	1/2 cup	87	105	17
Spaghetti, white	1/2 cup	110	70	20
Couscous	1/2 cup	88	78	17
Prunes	5 prunes	100	40	21
Potatoes, Russet, baked	1/2 medium	84	86	16
All Bran, Kellogg's	1/2 cup	80	31	13
Macaroni & Cheese, boxed	1 cup	490	193	49
Bread, whole grain wheat	1 slice	100	43	15
Apricots, dried	1/4 cup	110	40	24
Oatmeal cookie with raisins	Two 2 5/8 "	130	30	10
M&M Peanut	Half 8.27 oz	240	46	26
Fructose (fruit sugar)	1 tsp	15	4	4
Rice, white	1/2 cup	121	93	26
Bread, pumpernickel	1 slice	65	26	10
Pancake, plain	4 " pancake	86	38	11
Chocolate cake, with frosting	1/12 cake	158	43	22
Shortbread cookie	Three 1 5/8 "	120	24	8

FOOD	SIZE	CAL	GM	CARB
Blueberry muffin	1/2 large	260	64	29
Cranberry Sauce	1/4 cup	110	70	25
Bread, sourdough	1 slice	100	43	18
Waffle, plain	2 squares	109	38	12
Apple muffin	1/2 large	250	54	32
Banana bread	1 slice	196	60	32
Potato chips	1 oz	155	28	13
Corn tortilla	1 tortilla	52	24	9
Bread, seven grain	1 slice	100	38	18
Rye crispbread	3 crackers	112	29	12
Power Bar	1 bar	236	65	41
Bread, rye	1 slice	80	28	13
Pound cake	1/12 cake	109	28	14
Dates	1/4 cup	126	55	30
Pita	4 " pita	77	28	14
Twix cookie bar	1 cookie	120	24	16
Bread, wheat	1 slice	70	29	13
Chocolate bar	Half 2.6 oz	200	26	19
Corn bread	1 piece	188	60	28
Bread, white	1 slice	70	29	14
Mueslix, Kellogg's	2/3 cup	196	55	36
Doughnut, cake	One 4	297	71	32
Figs, dried	1/4 cup	93	37	21
Jams and jellies	1 tbsp	48	20	14
Corn chips	1 oz	147	28	16
Bagel	1/2 bagel	115	43	22
Soda crackers	4 crackers	104	24	12
Raisins	1/4 cup	130	50	29
English muffin	1/2 muffin	75	33	16

FOOD	SIZE	CAL	GM	CARB
Angel food cake	1/12 cake	72	28	16
Snickers bar	2/3 3.7 oz	170	35	20
Raisin Bran	1 cup	190	59	38
Bran Chex, Gen Mills	1 cup	166	51	35
Wheat thins	10 crackrs	90	20	12
Mini Wheats, frosted	1 cup	200	59	42
Whole grain bar, Kudos	1 bar	100	24	16
Popcorn, air popped	2 cups	62	16	10
Cupcake with frosting	1 cupcake	131	43	27
Special K, Kellogg's	1 cup	110	31	21
Cheerios, Gen Mills	1 cup	110	30	19
Vanilla wafers	3 cookies	84	18	12
Grapenuts, Post	1/2 cup	200	58	41
Total, General Mills	1 cup	133	40	27
Shredded Wheat	1 cup	170	49	34
Graham crackers	Four 2 1/2	120	28	20
Honey	1 tsp	16	5	5
Golden Grahams, General Mills	1 cup	160	41	33
Grape Nut Flakes, Post	1 cup	147	39	28
Table sugar	1 tsp	16	4	4
Coco Puffs, Gen Mills	1 cup	147	36	29
Pretzels	1 oz	108	28	22
Corn Flakes	1 cup	100	28	23
Crispy Rice	1 cup	96	26	23
Fruit roll ups	2 rolls	104	28	24

TABLE TWO: CALCULATION OF CARBOHYDRATE DENSITY AND GLYCEMIC DENSITY SCORE

GM = Total Grams, CARB = Carbohydrate Grams, CD = Carbohydrate Density, GI = Glycemic Index, GDS = Glycemic Density Score, "-" = Not Measurable
Note: Dry cereal listings are without milk. Table is in Carbohydrate Density (CD) order.

FOOD	GM	CARB	CD	GI	GDS
Spinach, cooked	180	2	.04	-	-
Cauliflower, cooked	124	2	.06	-	-
Lettuce, iceberg, raw	72	1	.06	-	-
Mushrooms, cooked	108	2	.07	-	-
Radishes, raw	116	2	.07	-	-
Cucumber, raw, sliced	119	2	.07	-	-
Celery, raw, chopped	101	2	.08	-	-
Lettuce, Romaine, raw	47	1	.09	-	-
Zucchini, cooked	180	4	.09	-	-
Brussels sprouts, cooked	156	4	.10	-	-
Cottage Cheese, low fat (1%)	113	3	.11	-	-
Cabbage, raw, shredded	70	2	.11	-	-
Cottage Cheese, plain	113	3	.11	-	-
Tomatoes, raw, chopped	180	5	.11	-	-
Cottage Cheese, reduced fat (2%)	113	3	.11	-	-
Carrots, raw, chopped	108	3	.11	47	.052
Cauliflower, raw	100	3	.12	-	-

FOOD	GM	CARB	CD	GI	GDS
Asparagus, cooked	180	6	.13	-	-
Green pepper, raw, chopped	149	5	.13	-	-
Tomatoes, cooked	240	8	.13	-	-
Spinach, raw	30	1	.13	-	-
Broccoli, cooked	156	6	.15	-	-
Green beans, cooked	135	5	.15	-	-
Milk, soy	262	10	.15	20	.030
Olives	44	2	.18	-	-
Cottage Cheese, nonfat	113	5	.18	-	-
Avocados	68	3	.18	-	-
Broccoli, raw	91	4	.18	-	-
Milk, whole	244	11	.18	27	.049
Snap peas, raw, whole	63	3	.19	-	-
Green onions, raw	100	5	.20	-	-
Milk, reduced fat (2%)	244	12	.20	28	.056
Milk, lowfat (1%)	244	12	.20	29	.058
Milk, nonfat	245	13	.21	30	.063
Beets, canned, drained	170	9	.21	64	.134
Strawberries	144	8	.22	40	.088
Yogurt, nonfat flavored, no sugar added	226	13	.23	30	.069
Soybeans, fresh	128	9	.28	18	.050
Grapefruit	230	16	.28	35	.098
Pudding, sugar free, nonfat milk	128	9	.28	44	.123
Milk, chocolate, nonfat, no sugar added	250	18	.29	24	.070
Watermelon	154	11	.29	72	.209

FOOD	GM	CARB	CD	GI	GDS
Yogurt, nonfat, plain	227	17	.30	28	.084
Cantaloupe	160	12	.30	65	.195
Kidney beans, canned	128	10	.31	28	.087
Oatmeal, one minute	279	22	.32	66	.211
Oatmeal, old fashioned	280	23	.33	42	.139
Peaches	154	13	.34	42	.143
Cream of Wheat, Nabisco	273	23	.34	66	.224
Apricots	165	15	.36	57	.205
Pears, in juice	125	12	.38	43	.163
Squash, winter, cooked	102	10	.39	0	.000
Peaches, in juice	124	12	.39	38	.148
Plumbs	165	17	.41	39	.160
Oranges	165	17	.41	42	.172
Milk, chocolate (1%)	250	26	.42	34	.143
Fruit cocktail, in juice	124	13	.42	55	.231
Potatoes, instant, milk/butter	105	11	.42	85	.357
Pinto beans, canned	120	13	.43	42	.181
Custard	141	15	.43	43	.185
Apples	110	12	.44	38	.167
Pineapple	155	17	.44	59	.260
Peaches, in light syrup	124	14	.45	52	.234
Apricots, in juice	124	14	.45	61	.275
Zoom or Wheatena	195	23	.47	25	.118
Kiwi	177	21	.47	53	.249
Lentils, cooked	99	12	.48	25	.120
Potatoes, french fried	100	12	.48	75	.360
Pears	140	17	.49	38	.186
Peas, cooked	98	12	.49	42	.206

FOOD	GM	CARB	CD	GI	GDS
Applesauce, no added sugar	126	16	.51	40	.204
Cherries, no pits	145	20	.55	22	.121
Lima beans, cooked	94	13	.55	32	.176
Mango	165	25	.61	42	.256
Sweet potato, cooked	151	23	.61	44	.268
Potatoes, mashed, milk/butter	105	16	.61	73	.445
Navy beans, cooked	91	14	.62	38	.236
Baked beans, canned	127	20	.63	48	.302
Potatoes, mashed with milk	105	17	.65	73	.475
Buckwheat	84	14	.67	54	.362
Corn	77	13	.68	54	.367
Grapes, Thompson, seedless	151	26	.69	46	.317
Potatoes, boiled	78	14	.72	56	.403
Peaches, in heavy syrup	127	23	.72	58	.418
Pudding	145	27	.74	44	.326
Potatoes, Russet, baked	86	16	.74	85	.629
Yogurt, fruit on bottom	245	46	.75	33	.248
Ice cream, premium	74	14	.76	37	.281
Banana	118	24	.81	50	.405
Peanuts	28	6	.86	1	.009
Rice, brown	98	21	.86	55	.473
Couscous	78	17	.87	65	.566
Fettucinne, egg	90	21	.93	40	.372
Yams, cooked	68	16	.94	37	.348
Pizza, thin, 14 "	123	29	.94	39	.367

FOOD	GM	CARB	CD	GI	GDS
Ice cream, nonfat, no sugar added	68	16	.94	50	.470
Barley, pearled	78	19	.97	25	.243
Ice cream, light, low fat	62	15	.97	37	.359
Macaroni & Cheese, boxed	193	49	1.02	64	.653
Spaghetti, whole wheat	70	18	1.03	37	.381
Ice cream, regular	66	18	1.09	38	.414
Rice, white	93	26	1.12	68	.762
Spaghetti, white	70	20	1.14	44	.502
Pancake, plain	38	11	1.16	67	.777
Spaghetti, protein enriched	70	21	1.20	26	.312
Waffle, plain	38	12	1.26	76	.958
Oatmeal cookie with raisins	30	10	1.33	55	.732
Shortbread cookie	24	8	1.33	64	.851
Bread, whole grain wheat	43	15	1.40	51	.714
Cranberry Sauce	70	25	1.43	60	.858
Corn tortilla	24	9	1.50	68	1.020
Bread, pumpernickel	26	10	1.54	50	.770
Rye crispbread	29	12	1.66	63	1.046
Bread, sourdough	43	18	1.67	54	.902
All Bran, Kellogg's	31	13	1.68	38	.638
Bread, wheat	29	13	1.79	70	1.253
Doughnut, cake	71	32	1.80	76	1.368
Blueberry muffin	64	29	1.81	47	.851
Cashews	28	13	1.86	3	.056
Potato chips	28	13	1.86	54	1.004

FOOD	GM	CARB	CD	GI	GDS
Bread, rye	28	13	1.86	58	1.079
Corn bread	60	28	1.87	68	1.272
Bread, seven grain	38	18	1.89	55	1.040
Bread, white	29	14	1.93	70	1.351
English muffin	33	16	1.94	77	1.494
Pound cake	28	14	2.00	54	1.080
Pita	28	14	2.00	57	1.140
Soda crackers	24	12	2.00	74	1.480
Chocolate cake, frosting	43	22	2.05	38	.779
Bagel	43	22	2.05	72	1.476
Prunes	40	21	2.10	29	.609
Banana bread	60	32	2.13	47	1.001
Dates	55	30	2.18	50	1.090
M&M Peanut	46	26	2.26	33	.746
Figs, dried	37	21	2.27	61	1.385
Corn chips	28	16	2.29	63	1.443
Angel food cake	28	16	2.29	67	1.534
Snickers bar	35	20	2.29	68	1.557
Raisins	50	29	2.32	64	1.485
Apple muffin	54	32	2.37	41	.972
Apricots, dried	40	24	2.40	30	.720
Wheat thins	20	12	2.40	67	1.608
Popcorn, air popped	16	10	2.50	72	1.800
Cupcake with frosting	43	27	2.51	73	1.832
Power Bar	65	41	2.52	42	1.058
Cheerios, General Mills	30	19	2.53	74	1.872
Raisin Bran	59	38	2.58	61	1.574
Mueslix, Kellogg's	55	36	2.62	52	1.362
Twix cookie bar	24	16	2.67	44	1.175

FOOD	GM	CARB	CD	GI	GDS
Whole grain bar, Kudos	24	16	2.67	62	1.655
Vanilla wafers	18	12	2.67	74	1.976
Total, General Mills	40	27	2.70	76	2.052
Special K, Kellogg's	31	21	2.71	69	1.870
Bran Chex, General Mills	51	35	2.75	58	1.595
Shredded Wheat	49	34	2.78	75	2.085
Jams and jellies	20	14	2.80	51	1.428
Grapenuts, Post	58	41	2.83	71	2.009
Mini Wheats, frosted	59	42	2.85	58	1.653
Graham crackers	28	20	2.86	74	2.116
Grape Nut Flakes, Post	39	28	2.87	80	2.296
Chocolate bar	26	19	2.92	43	1.256
Pretzels	28	22	3.14	83	2.606
Golden Grahams, General Mills	41	33	3.22	71	2.286
Coco Puffs, Gen Mills	36	29	3.22	77	2.479
Corn Flakes	28	23	3.29	81	2.665
Fruit roll ups	28	24	3.43	99	3.396
Crispy Rice	26	23	3.54	82	2.903
Fructose (fruit sugar)	4	4	4.00	19	.760
Honey	5	5	4.00	55	2.200
Table sugar	4	4	4.00	61	2.440

TABLE THREE:
100 GRAM AND CEREAL LISTS

LIST OF FOODS IN 100 GRAM SERVINGS

Cal = Total Calories, Net Carb = Carbohydrate Grams without fiber, GI = Glycemic Index, GL = Glycemic Load, GDS = Glycemic Density Score "-" = Not Measurable

Note: Dry cereal listings are without milk. Table is in Glycemic Density (GDS) order.

Food	Cal	Net Carb	GI	GL	GDS
Very Low Glycemic Density					
Spinach, cooked	23	1.1	-	-	-
Cauliflower, cooked	23	1.6	-	-	-
Lettuce, iceberg, raw	14	1.4	-	-	-
Mushrooms, cooked	26	1.9	-	-	-
Radishes, raw	16	1.7	-	-	-
Cucumber, raw, sliced	12	1.7	-	-	-
Celery, raw, chopped	16	2.0	-	-	-
Lettuce, Romaine, raw	17	2.1	-	-	-
Zucchini, cooked	16	2.2	-	-	-
Brussels sprouts, cooked	36	2.6	-	-	-
Cottage Cheese, plain	102	2.6	-	-	-
Cottage Cheese, reduced fat (2%)	89	2.6	-	-	-
Cottage Cheese, low fat (1%)	72	2.6	-	-	-
Cabbage, raw, shredded	26	2.9	-	-	-
Tomatoes, raw, chopped	18	2.8	-	-	-
Cauliflower, raw	25	3.0	-	-	-

Food	Cal	Net Carb	GI	GL	GDS
Very Low Glycemic Density (Continued)					
Asparagus, cooked	22	3.4	-	-	-
Green pepper, raw, chopped	20	3.4	-	-	-
Tomatoes, cooked	18	3.4	-	-	-
Spinach, raw	23	3.3	-	-	-
Broccoli, cooked	35	3.8	-	-	-
Green beans, cooked	28	3.7	-	-	-
Olives	114	4.5	-	-	-
Avocados	168	4.4	-	-	-
Broccoli, raw	34	4.4	-	-	-
Cottage Cheese, nonfat	70	4.4	-	-	-
Snap peas, raw, whole	41	4.8	-	-	-
Green onions, raw	32	5.0	-	-	-
Squash, winter, cooked	55	9.8	-	-	-
Peanuts	593	21.4	1	.2	.01
Milk, soy	43	3.8	20	1	.03
Milk, whole	60	4.5	27	1	.05
Soybeans, fresh	147	7.0	18	1	.05
Carrots, raw, chopped	48	2.8	47	1	.05
Milk, reduced fat (2%)	50	4.9	28	1	.06
Milk, lowfat (1%)	42	4.9	29	1	.06
Cashews	582	46.4	3	1	.06
Milk, nonfat	37	5.3	30	1	.06
Milk, chocolate, nonfat, no sugar added	52	7.2	24	2	.07
Yogurt, nonfat flavored, no sugar added	44	5.7	30	2	.07
Yogurt, nonfat, plain	56	7.5	28	2	.08
Kidney beans, canned	84	7.8	28	2	.09

Food	Cal	Net Carb	GI	GL	GDS
Very Low Glycemic Density (Continued)					
Strawberries	32	5.5	40	2	.09
Grapefruit	32	6.9	35	2	.10
Zoom or Wheatena hot cereal	71	11.7	25	3	.12
Lentils, cooked	116	12.1	25	3	.12
Cherries, no pits	63	13.8	22	3	.12
Pudding, sugar free nonfat milk	55	7.0	44	3	.12
Beets, canned, drained	31	5.3	64	3	.14
Oatmeal, old fashioned	54	8.3	42	3	.14
Milk, chocolate (1%)	63	10.4	34	4	.14
Peaches	39	8.5	42	4	.14
Peaches, in juice	41	9.7	38	4	.15
Plumbs	46	10.4	39	4	.16
Pears, in juice	48	9.6	43	4	.16
Apples	52	10.9	38	4	.17
Oranges	49	10.4	42	4	.17
Lima beans, cooked	114	13.8	32	4	.18
Pinto beans, canned	85	10.8	42	5	.18
Custard	105	10.7	43	5	.19
Pears	58	12.1	38	5	.19
Cantaloupe	34	7.6	65	5	.20
Applesauce, no added sugar	55	12.6	40	5	.20
Low Glycemic Density					
Apricots	48	9.2	57	5	.21
Peas, cooked	118	12.2	42	5	.21
Watermelon	30	7.2	72	5	.21
Oatmeal, one minute	50	7.9	66	5	.21

Food	Cal	Net Carb	GI	GL	GDS
Low Glycemic Density (Continued)					
Cream of Wheat, Nabisco	44	8.5	66	6	.22
Fruit cocktail, in juice	49	10.5	55	6	.23
Peaches, in light syrup	49	11.3	52	6	.23
Navy beans, cooked	141	15.4	38	6	.24
Barley, pearled	123	24.3	25	6	.24
Yogurt, fruit on bottom	94	18.9	22	6	.25
Kiwi	60	11.8	53	6	.25
Mango	65	15.3	42	6	.26
Pineapple	48	11.1	59	7	.26
Sweet potato, cooked	76	15.2	44	7	.27
Apricots, in juice	49	11.3	61	7	.28
Ice cream, premium	254	18.9	37	7	.28
Baked beans, canned	95	15.8	48	8	.30
Spaghetti, protein enriched	164	30.0	26	8	.31
Grapes, Thompson, seedless	69	17.2	46	8	.32
Pudding	93	18.6	44	8	.33
Yams, cooked	116	23.5	37	9	.35
Potatoes, instant milk/butter	97	10.5	85	9	.36
Ice cream, light, low fat	177	24.2	37	9	.36
Potatoes, french fried	134	12.0	75	9	.36
Buckwheat	93	16.7	54	9	.36
Corn	86	16.9	54	9	.37
Pizza, thin, 14 "	247	23.5	39	9	.37
Fettucinne, egg	122	23.3	40	9	.37
Spaghetti, whole wheat	124	25.7	37	10	.38

Food	Cal	Net Carb	GI	GL	GDS
Moderate Glycemic Density					
Potatoes, boiled	87	17.9	56	10	.40
Banana	89	20.4	50	10	.41
Ice cream, regular	217	27.4	38	10	.41
Peaches, in heavy syrup	79	18.2	58	11	.42
Potatoes, mashed milk/butter	113	15.2	73	11	.45
Ice cream, nonfat, no sugar added	135	23.5	50	12	.47
Rice, brown	111	21.4	55	12	.47
Potatoes, mashed with milk	83	16.2	73	12	.48
Spaghetti, white	157	28.6	44	13	.50
Couscous	113	21.8	65	14	.57
High Glycemic Density					
Prunes	250	52.5	29	15	.61
Potatoes, Russet, baked	97	18.6	85	16	.63
All Bran, Kellogg's	258	42.0	38	16	.64
Macaroni & Cheese, boxed	255	25.5	64	16	.65
Bread, whole grain wheat	233	35.0	51	18	.71
Apricots, dried	275	60.0	30	18	.72
Oatmeal cookie with raisins	433	33.3	55	18	.73
M&M Peanut	521	56.4	33	19	.75
Fructose (fruit sugar)	375	100.0	19	19	.76
Rice, white	131	28.1	68	19	.76
Bread, pumpernickel	250	38.5	50	19	.77
Pancake, plain	226	28.9	67	19	.78
Chocolate cake, frosting	368	51.3	38	19	.78

Food	Cal	Net Carb	GI	GL	GDS
High Glycemic Density (Continued)					
Shortbread cookie	500	33.4	64	21	.85
Blueberry muffin	406	45.2	47	21	.85
Cranberry Sauce	157	35.8	60	21	.86
Bread, sourdough	233	41.9	54	23	.90
Waffle, plain	287	31.6	76	24	.96
Apple muffin	463	59.2	41	24	.97
Banana bread	327	53.4	47	25	1.00
Potato chips	553	46.4	54	25	1.00
Corn tortilla	217	37.5	68	26	1.02
Bread, seven grain	263	47.3	55	26	1.04
Rye crispbread	386	41.4	63	26	1.05
Power Bar	363	63.1	42	26	1.06
Bread, rye	286	46.4	58	27	1.08
Pound cake	389	50.0	54	27	1.08
Dates	229	54.6	50	27	1.09
Pita	275	50.0	57	28	1.14
Twix cookie bar	500	66.7	44	29	1.18
Bread, wheat	242	44.9	70	31	1.25
Chocolate bar	770	73.2	43	31	1.26
Corn bread	314	46.8	68	32	1.27
Bread, white	242	48.3	70	33	1.35
Mueslix, Kellogg's	357	65.5	52	34	1.36
Doughnut, cake	419	45.1	76	34	1.37
Figs, dried	251	56.7	61	35	1.39
Jams and jellies	240	70.0	51	36	1.43
Corn chips	525	57.1	63	36	1.44
Bagel	268	51.3	72	37	1.48

Food	Cal	Net Carb	GI	GL	GDS
High Glycemic Density (Continued)					
Soda crackers	434	50.0	74	37	1.48
Raisins	260	58.0	64	37	1.49
English muffin	227	48.5	77	37	1.49
Angel food cake	257	57.1	67	38	1.53
Snickers bar	486	57.2	68	39	1.56
Raisin Bran	321	64.2	61	39	1.57
Bran Chex, General Mills	325	68.6	58	40	1.60
Wheat thins	450	60.0	67	40	1.61
Mini Wheats, Frosted	338	71.0	58	41	1.65
Whole Grain Bar, Kudos	417	66.7	62	41	1,66
Popcorn, air popped	388	62.5	72	45	1.80
Cupcake with frosting	305	62.9	73	46	1.83
Special K, Kellogg's	355	67.8	69	47	1.87
Cheerios, General Mills	366	63.3	74	47	1.87
Vanilla wafers	467	66.7	74	49	1.98
Grapenuts, Post	344	70.5	71	50	2.01
Total, General Mills	333	67.5	76	51	2.05
Shredded Wheat	347	69.4	75	52	2.09
Graham crackers	428	71.4	74	53	2.12
Honey	320		55	55	2.20
Golden Grahams, Gen Mills	390	80.5	71	57	2.29
Grape Nut Flakes, Post	376	71.7	80	57	2.30
Table sugar	400		61	61	2.44
Coco Puffs, General Mills	409	80.6	77	62	2.48
Pretzels	386	78.5	83	65	2.61
Corn Flakes	357	82.1	81	66	2.67
Crispy Rice	370	88.6	82	73	2.90
Fruit roll ups	371	85.7	99	85	3.40

BREAKFAST CEREAL COMPARISON

(One Ounce Of Cereal Without Milk =28.35 gm)

Food	Cal	Net Carb	GI	GL	GDS
All Bran, Kellogg's	74	12.0	38	4	.64
Mueslix, Kellogg's	103	18.9	52	10	1.37
Raisin Bran	91	18.2	61	11	1.57
Bran Chex, General Mills	93	19.6	58	11	1.60
Mini Wheats, frosted	96	20.2	58	12	1.65
Cheerios, General Mills	105	18.1	74	13	1.87
Special K, Kellogg's	100	19.1	69	13	1.87
Total, General Mills	94	19.2	76	15	2.05
Shredded Wheat	99	19.7	75	15	2.09
Grape Nut Flakes, Post	107	19.7	80	16	2.22
Grapenuts, Post	112	23.0	71	16	2.29
Golden Grahams, Gen Mills	110	22.8	71	16	2.29
Coco Puffs, General Mills	116	22.9	77	17	2.48
Corn Flakes	101	23.2	81	19	2.67
Crispy Rice	105	25.1	82	21	2.90

BREAKFAST CEREAL COMPARISON

One Ounce of Cereal with 1/2 Cup Nonfat Milk = 152 gm)

Food	Cal	Net Carb	GI	GL	GDS
All Bran, Kellogg's	119	18.0	33	6	.16
Mueslix, Kellogg's	148	24.9	48	12	.32
Raisin Bran	136	24.2	54	13	.35
Bran Chex, General Mills	138	25.6	51	13	.34
Mini Wheats, frosted	141	26.2	53	14	.37
Cheerios, General Mills	150	24.1	62	15	.39
Special K, Kellogg's	145	25.1	60	15	.40
Total, General Mills	139	25.2	67	17	.45
Shredded Wheat	144	25.7	66	17	.45
Grape Nut Flakes, Post	152	25.7	70	18	.48
Grapenuts, Post	157	29.0	62	18	.47
Golden Grahams, Gen Mills	155	28.8	63	18	.48
Coco Puffs, General Mills	161	28.9	66	19	.51
Corn Flakes	146	29.2	72	21	.55
Crispy Rice	150	31.1	74	23	.61

TABLE FOUR: GLYCEMIC-DENSITY ENERGY-DENSITY COMPARISON

Note: Dry cereal listings are without milk

Food	ED #	ED Level	Reason
Very Low Glycemic Density			
Spinach, cooked	.23	Very Low	
Cauliflower, cooked	.23	Very Low	
Lettuce, iceberg, raw	.14	Very Low	
Mushrooms, cooked	.26	Very Low	
Radishes, raw	.16	Very Low	
Cucumber, raw, sliced	.12	Very Low	
Celery, raw, chopped	.16	Very Low	
Lettuce, Romaine, raw	.17	Very Low	
Zucchini, cooked	.16	Very Low	
Brussels sprouts, cooked	.35	Very Low	
Cottage Cheese, plain	1.03	Low	Protein Fat
Cottage Cheese, reduced fat (2%)	.88	Low	Protein Fat
Cottage Cheese, low fat (1%)	.72	Low	Protein Fat
Cabbage, raw, shredded	.25	Very Low	

Food	ED #	ED Level	Reason
Very Low Glycemic Density (Continued)			
Tomatoes, raw, chopped	.18	Very Low	
Cauliflower, raw	.25	Very Low	
Asparagus, cooked	.22	Very Low	
Green pepper, raw, chopped	.20	Very Low	
Tomatoes, cooked	.18	Very Low	
Spinach, raw	.23	Very Low	
Broccoli, cooked	.34	Very Low	
Green beans, cooked	.28	Very Low	
Olives	1.14	Low	Fat
Avocados	1.67	Moderate	Fat
Broccoli, raw		Very Low	
Cottage Cheese, nonfat	.71	Low	Protein
Snap peas, raw, whole	.41	Very Low	
Green onions, raw	.32	Very Low	
Squash, winter, cooked	.55	Very Low	
Peanuts	5.83	High	Fat
Milk, soy	.43	Very Low	
Milk, whole	.56	Very Low	
Soybeans, fresh	1.47	Low	Protein Fat
Carrots, raw, chopped	.48	Very Low	
Milk, reduced fat (2%)	.46	Very Low	

Food	ED #	ED Level	Reason
Very Low Glycemic Density (Continued)			
Milk, lowfat (1%)	.39	Very Low	
Cashews	5.72	High	Fat
Milk, nonfat	.34	Very Low	
Milk, chocolate, nonfat, no sugar added	.48	Very Low	
Yogurt, nonfat flavored, no sugar added	.44	Very Low	
Yogurt, nonfat, plain	.46	Very Low	
Kidney beans, canned	.84	Low	Protein
Strawberries	.32	Very Low	
Grapefruit	.32	Very Low	
Zoom or Wheatena hot cereal	.72	Low	Protein
Lentils, cooked	1.16	Low	Protein
Cherries, no pits	.63	Low	Protein
Pudding, sugar free nonfat milk	.55	Very Low	
Milk, chocolate (1%)	.57	Very Low	
Beets, canned, drained	.31	Very Low	
Oatmeal, old fashioned	.39	Very Low	
Peaches	.39	Very Low	
Peaches, in juice	.40	Very Low	
Plumbs	.46	Very Low	

Food	ED #	ED Level	Reason
Very Low Glycemic Density (Continued)			
Pears, in juice	.48	Very Low	
Apples	.52	Very Low	
Oranges	.49	Very Low	
Lima beans, cooked	1.15	Low	Protein
Pinto beans, canned	.86	Low	Protein
Custard	1.05	Low	Protein Fat
Pears	.58	Very Low	
Cantaloupe	.34	Very Low	
Applesauce, no added sugar	.56	Very Low	
Low Glycemic Density			
Apricots	.48	Very Low	Glycemic
Peas, cooked	1.18	Low	
Watermelon	.30	Very Low	Glycemic
Oatmeal, one minute	.50	Very Low	Glycemic
Cream of Wheat, Nabisco	.44	Very Low	Glycemic
Fruit cocktail, in juice	.48	Very Low	Glycemic
Peaches, in light syrup	.48	Very Low	Glycemic
Navy beans, cooked	1.40	Low	
Barley, pearled	1.23	Low	
Yogurt, fruit on bottom	.94	Low	

Food	ED #	ED Level	Reason
Low Glycemic Density (Continued)			
Kiwi	.61	Low	
Mango	.65	Low	
Pineapple	.48	Very Low	Glycemic
Sweet potato, cooked	.76	Low	
Apricots, in juice	.48	Low	
Ice cream, premium	2.54	Moderate	Fat
Baked beans, canned	.94	Low	
Spaghetti, protein enriched	1.64	Moderate	Protein
Grapes, Thompson, seedless	.69	Low	
Pudding	.93	Low	
Yams, cooked	1.16	Low	
Potatoes, instant milk/butter	.97	Low	
Ice cream, light, low fat	1.77	Moderate	Fat
Potatoes, french fried	1.34	Low	
Buckwheat	.92	Low	
Corn	,86	Low	
Pizza, thin, 14 "	2.48	Moderate	Protein Fat
Fettucinne, egg	1.22	Low	
Spaghetti, whole wheat	1.24	Low	

Food	ED #	ED Level	Reason
Moderate Glycemic Density			
Potatoes, boiled	.87	Low	
Banana	.89	Low	Glycemic
Ice cream, regular	2.17	Moderate	
Peaches, in heavy syrup	.79	Low	Glycemic
Potatoes, mashed milk/butter	1.13	Low	Glycemic
Ice cream, nonfat, no sugar added	1.35	Low	Glycemic
Rice, brown	1.12	Low	Glycemic
Potatoes, mashed with milk	.83	Low	Glycemic
Spaghetti, white	1.57	Moderate	
Couscous	1.12	Low	Glycemic
High Glycemic Density			
Prunes	2.5	Moderate	Glycemic
Potatoes, Russet, baked	.97	Low	Glycemic
All Bran, Kellogg's	2.58	Moderate	Glycemic
Macaroni & Cheese, boxed	2.54	Moderate	Glycemic
Bread, whole grain wheat	2.33	Moderate	Glycemic
Apricots, dried	2.75	Moderate	Glycemic
Oatmeal cookie with raisins	4.33	High	
M&M Peanut	5.22	High	
Fructose (fruit sugar)	3.57	Moderate	Glycemic
Rice, white	1.2	Low	Glycemic
Bread, pumpernickel	2.5	Moderate	Glycemic

Food	ED #	ED Level	Reason
High Glycemic Density (Continued)			
Pancake, plain	2.26	Moderate	Glycemic
Chocolate cake, with frosting	3.67	Moderate	Glycemic
Shortbread cookie	5.00	High	
Blueberry muffin	4.06	High	
Cranberry Sauce	1.57	Moderate	Glycemic
Bread, sourdough	2.33	Moderate	Glycemic
Waffle, plain	2.87	Moderate	Glycemic
Apple muffin	3.9	Moderate	Glycemic
Banana bread	3.27	Moderate	Glycemic
Potato chips	5.44	High	
Corn tortilla	2.16	Moderate	Glycemic
Bread, seven grain	2.63	Moderate	Glycemic
Rye crispbread	3.82	Moderate	Glycemic
Power Bar	3.63	Moderate	Glycemic
Bread, rye	2.86	Moderate	Glycemic
Pound cake	3.89	Moderate	Glycemic
Dates	2.86	Moderate	Glycemic
Pita	2.75	Moderate	Glycemic
Twix cookie bar	5.00	High	
Bread, wheat	2.41	Moderate	Glycemic
Chocolate bar	5.56	High	
Corn bread	3.13	Moderate	Glycemic
Bread, white	2.41	Moderate	Glycemic
Mueslix, Kellogg's	3.56	Moderate	Glycemic
Doughnut, cake	4.18	High	
Figs, dried	2.51	Moderate	Glycemic
Jams and jellies	2.40	Moderate	Glycemic
Corn chips	5.16	High	

Food	ED #	ED Level	Reason
High Glycemic Density (Continued)			
Bagel	2.70	Moderate	Glycemic
Soda crackers	4.33	High	
Raisins	3.25	Moderate	Glycemic
English muffin	2.27	Moderate	Glycemic
Angel food cake	2.57	Moderate	Glycemic
Snickers bar	4.86	High	
Raisin Bran	3.22	Moderate	Glycemic
Bran Chex, Gen Mills	3.26	Moderate	Glycemic
Wheat thins	4.5	High	
Mini Wheats, frosted	3.39	Moderate	Glycemic
Whole Grain Bar	4.17	High	
Popcorn, air popped	3.88	Moderate	Glycemic
Cupcake with frosting	3.05	Moderate	Glycemic
Special K, Kellogg's	3.55	Moderate	Glycemic
Cheerios, Gen Mills	3.67	Moderate	Glycemic
Vanilla wafers	4.67	High	
Grapenuts, Post	3.45	Moderate	Glycemic
Total, General Mills	3.32	Moderate	Glycemic
Shredded Wheat	3.47	Moderate	Glycemic
Graham crackers	4.29	High	
Honey	3.05	Moderate	Glycemic
Golden Grahams	3.87	Moderate	Glycemic
Grape Nut Flakes	3.79	Moderate	Glycemic
Table sugar	3.81	Moderate	Glycemic
Coco Puffs, Gen Mills	4.07	High	
Pretzels	3.80	Moderate	Glycemic
Corn Flakes	3.57	Moderate	Glycemic
Crispy Rice	3.69	Moderate	Glycemic
Fruit roll ups	3.71	Moderate	Glycemic

TABLE FIVE: LIST OF FOODS IN FOOD EXCHANGE CALORIES

Note: Dry cereal listings are based upon 1/2 cup of milk for each ounce of cereal.

Low Carbohydrate Vegetables (25 Calories)

FOOD	SIZE	MEAS	GM	CARB	GDS
Very Low Glycemic Density					
Cauliflower, raw	1.00	Cup	100	3.0	-
Lettuce, iceberg, raw	2.50	Cup	180	2.5	-
Green onions, raw	.78	Cup	78	3.9	-
Cauliflower, cooked	.89	Cup	110	1.8	-
Spinach, raw	3.57	Cup	107	3.6	-
Radishes, raw	1.32	Cup	153	2.6	-
Cabbage, raw, shredded	1.39	Cup	97	2.8	-
Lettuce, Romaine, raw	3.13	Cup	147	3.1	-
Brussels sprouts, cooked	.45	Cup	70	1.8	-
Zucchini, cooked	.89	Cup	160	3.6	-
Tomatoes, raw, chopped	.78	Cup	140	3.9	-
Cucumber, raw, sliced	1.79	Cup	213	3.6	-

Low Carbohydrate Vegetables (Continued)

FOOD	SIZE	MEAS	GM	CARB	GDS
Very Low Glycemic Density (Continued)					
Celery, raw, chopped	1.56	Cup	158	3.1	-
Mushrooms, cooked	.89	Cup	96	1.8	-
Broccoli, raw	.81	Cup	74	3.2	-
Green pepper, raw, chopped	.83	Cup	124	4.2	-
Green beans, cooked	.66	Cup	89	3.3	-
Asparagus, cooked	.63	Cup	113	3.8	-
Broccoli, cooked	.46	Cup	72	2.8	-
Tomatoes, cooked	.58	Cup	139	4.6	-
Snap peas, raw, whole	.96	Cup	60	2.9	-
Spinach, cooked	.61	Cup	110	1.2	-

Fruits (60 Calories)

FOOD	SIZE	MEAS	GM	CARB	GDS
Very Low Glycemic Density					
Strawberries	1.30	Cup	187	10.4	.09
Grapefruit	.81	Cup	186	13.0	.10
Cherries, no pits	.66	Cup	96	13.2	.12
Peaches	1.00	Cup	154	13.0	.14
Peaches, in juice	.60	Cup	149	14.4	.15
Pears, in juice	.50	Cup	125	12.0	.16

Fruits (Continued)

FOOD	SIZE	MEAS	GM	CARB	GDS
Very Low Glycemic Density (Continued)					
Plumbs	.79	Cup	130	13.4	.16
Apples	1.05	Cup	116	12.6	.17
Oranges	.74	Cup	122	12.6	.17
Pears	.74	Cup	104	12.6	.19
Cantaloupe	1.11	Cup	178	13.3	.20
Applesauce, no added sugar	.43	Cup	108	13.8	.20
Low Glycemic Density					
Apricots	.76	Cup	125	11.4	.21
Watermelon	1.30	Cup	200	14.3	.21
Fruit cocktail, in juice	.50	Cup	124	13.0	.23
Peaches, in light syrup	.50	Cup	124	14.0	.23
Kiwi	.56	Cup	99	11.8	.25
Mango	.56	Cup	92	14.0	.26
Pinneapple	.81	Cup	126	13.8	.26
Apricots, in juice	.50	Cup	124	14.0	.28
Grapes, Thompson, seedless	.29	Cup	88	15.1	.32
Moderate Glycemic Density					
Banana	3.99	Inch	67	13.7	.41
Peaches, in heavy syrup	.30	Cup	76	13.8	.42
High Glycemic Density					
Prunes	3.00	Prunes	24	12.6	.61
Apricots, dried	.14	Cup	22	13.2	.72

Fruits (Continued)

FOOD	SIZE	MEAS	GM	CARB	GDS
High Glycemic Density (Continued)					
Cranberry Sauce	.14	Cup	39	13.8	.86
Dates	.12	Cup	26	14.4	1.09
Figs, dried	.16	Cup	24	13.7	1.39
Raisins	.12	Cup	23	13.3	1.49

Grains and Starchy Vegetables (80 Calories)

FOOD	SIZE	MEAS	GM	CARB	GDS
Very Low Glycemic Density					
Squash, winter, cooked	.72	Cup	146	14.3	-
Carrots, raw, chopped	1.54	Cup	166	4.6	.05
Zoom Wheatena	.38	Cup	111	13.1	.12
Oatmeal, old fashioned	.53	Cup	148	12.2	.14
Beets	1.51	Cup	257	13.6	.14
All Bran	.50	Cup	154	19	.18
Low Glycemic Density					
Oatmeal, 1 min	.57	Cup	159	12.5	.21
Cream of Wheat, Nabisco	.67	Cup	183	15.4	.22
Barley, pearled	.42	Cup	65	15.8	.24
Sweet Potatoes	.70	Medium	106	16.1	.27
Spaghetti, protein enriched	.35	Cup	49	14.7	.31
Raisin Bran with milk	.42	Cup	148	22.0	.31

Grains and Starchy Vegetables (Continued)

FOOD	SIZE	MEAS	GM	CARB	GDS
Low Glycemic Density (Continued)					
Mueslix, Kellogg's with milk	.27	Cup	146	20.8	.32
Mini Wheats, frosted with milk	.40	Cup	147	22.8	.32
Bran Chex, General Mills with milk	.48	Cup	147	22.8	.32
Special K, Kellogg's with milk	.73	Cup	146	21.3	.33
Cheerios, General Mills with milk	.73	Cup	145	19.9	.34
Yams, cooked	.51	Cup	69	16.2	.35
Potatoes, french fried	6.00	Fries	60	7.2	.36
Buckwheat	.52	Cup	87	14.4	.36
Pizza, thin, 14 "	.26	Piece	32	7.5	.37
Corn	.61	Cup	93	15.7	.37
Fettucinne, egg	.37	Cup	66	15.3	.37
Grapenuts, Post with milk	.20	Cup	146	22.4	.37
Golden Grahams, General Mills with milk	.50	Cup	144	22.5	.38
Spaghetti, whole wheat	.46	Cup	64	16.6	.38

Grains and Starchy Vegetables (Continued)

FOOD	SIZE	MEAS	GM	CARB	GDS
Low Glycemic Density (Continued)					
Coco Puffs with milk	.54	Cup	142	21.7	.39
Grape Nut Flakes, Post with milk	.54	Cup	144	21.1	.39
Shredded Wheat with milk	.47	Cup	146	22.0	.39
Potatoes, boiled	.59	Cup	92	16.5	.40
Total, General Mills with milk	.60	Cup	147	22.2	.40
Moderate Glycemic Density					
Potatoes, mashed with milk, butter	.34	Cup	70	10.7	.45
Rice, brown	.37	Cup	72	15.3	.47
Corn Flakes with milk	.80	Cup	145	24.4	.47
Potatoes, mashed with milk	.46	Cup	97	15.6	.48
Crispy Rice with milk	.83	Cake	145	25.1	.49
Spaghetti, white	.37	Cup	51	14.6	.50
Couscous	.46	Cup	71	15.5	.57
High Glycemic Density					
Potatoes, Russet, baked	.48	Medium	82	15.2	.63

Grains and Starchy Vegetables (Continued)

FOOD	SIZE	MEAS	GM	CARB	GDS
High Glycemic Density (Continued)					
Macaroni & Cheese, boxed	.16	Cup	31	7.8	.65
Bread, whole grain wheat	.80	Slice	34	12.0	.71
Rice, white	.33	Cup	61	17.2	.76
Bread, pumpernickel	1.23	Slice	32	12.3	.77
Pancake, plain	.93	4 in cake	35	10.2	.78
Waffle, plain	1.46	Squares	28	8.8	.96
Corn tortilla	1.54	Tortilla	37	13.9	1.02
Bread, 7 grain	.80	Slice	30	14.4	1.04
Pita	1.04	4 " pita	29	14.6	1.14
Corn bread	.43	Piece	26	12.0	1.27
Bread, white	1.14	Slice	33	16.0	1.35
Bread, wheat	1.14	Slice	33	14.8	1.25
Bagel	.35	Bagel	30	15.4	1.48
English muffin	.54	Large	35	17.1	1.49

Protein Foods (110 Calories - Mayo Clinic Recommendation)

FOOD	SIZE	MEAS	GM	CARB	GDS
Very Low Glycemic Density					
Cottage Cheese, plain	.48	Cup	107	2.9	-
Cottage Cheese, reduced fat (2%)	.55	Cup	123	3.3	-
Cottage Cheese, nonfat	.69	Cup	156	6.9	-

Protein Foods (Continued)

FOOD	SIZE	MEAS	GM	CARB	GDS
Very Low Glycemic Density (Continued)					
Cottage Cheese, low fat (1%)	.67	Cup	151	4.0	-
Milk, soy	.98	Cup	257	9.8	.03
Soybeans, fresh	.30	Cup	76	5.3	.05
Milk, whole	.75	Cup	183	8.3	.05
Milk, reduced fat (2%)	.90	Cup	220	10.8	.06
Milk, lowfat (1%)	1.08	Cup	264	13.0	.06
Milk, nonfat	1.21	Cup	295	15.7	.06
Milk, chocolate, nonfat, no sugar added	.85	Cup	213	15.3	.07
Yogurt, nonfat flavored, no sugar added	1.10	Cup	249	14.3	.07
Yogurt, nonfat, plain	.87	Cup	197	14.8	.08
Kidney beans, canned	.51	Cup	131	10.2	.09
Lentils, cooked	.48	Cup	95	11.5	.12
Milk, chocolate (1%)	.70	Cup	175	18.2	.14
Pinto beans, canned	.54	Cup	128	13.9	.18
Lima beans, cooked	.51	Cup	96	13.3	.18

Protein Foods (Continued)

FOOD	SIZE	MEAS	GM	CARB	GDS
Low GlycemicDensity					
Peas, cooked	.48	Cup	93	11.4	.21
Navy beans, cooked	.43	Cup	78	12.0	.24
Yogurt, fruit on bottom	.48	Cup	118	22.1	.25
Baked beans, canned	.46	Cup	117	18.4	.30

Sugars (16 Calories)

FOOD	SIZE	MEAS	GM	CARB	GDS
High Glycemic Density					
Fructose (fruit sugar)	1.07	Tsp	4	4.3	.76
Jams and jellies	.33	Tbsp	7	4.6	1.43
Table sugar	1.00	Tsp	4	4.0	2.44
Honey	1.00	Tsp	5	5.0	2.20

Fats (45 Calories)

FOOD	SIZE	MEAS	GM	CARB	GDS
Very Low Glycemic Density					
Avocados	.20	Fruit	27	1.2	-
Olives	9.00	Large	40	1.8	-
Peanuts	8.10	Nuts	8	1.6	.01
Cashews	5.04	Nuts	8	3.6	.06

Indulgences (100 Calories-My Personal Choice)

FOOD	SIZE	MEAS	GM	CARB	GDS
Very Low Glycemic Density					
Pudding, sugar free nonfat milk	.72	Cup	183	12.9	.12
Custard	.34	Cup	96	10.2	.19
Low Glycemic Density					
Ice cream, premium	.27	Cup	39	7.4	.28
Pudding	.37	Cup	107	20.0	.33
Ice cream, light, low fat	.46	Cup	56	13.7	.36
Moderate Glycemic Density					
Ice cream, regular	.35	Cup	46	12.6	.41
Ice cream, nonfat, no sugar added	.55	Cup	74	17.4	.47
High Glycemic Density					
Oatmeal cookie with raisins	1.54	2 5/8 in	23	7.7	.73
M&M Peanut	.21	8.27 oz	19	10.9	.75
Chocolate cake, frosting	.05	Cup	27	13.9	.78
Shortbread cookie	2.49	1 5/8 in	20	6.6	.85
Blueberry muffin	.19	Large	24	11.0	.85
Apple muffin	.20	Large	22	12.8	.97
Potato chips	.65	Oz	18	8.5	1.00
Banana bread	.51	Slice	31	16.3	1.00

Indulgences (Continued)

FOOD	SIZE	MEAS	GM	CARB	GDS
High Glycemic Density (Continued)					
Rye crispbread	2.67	Crckrs	26	10.7	1.05
Power Bar	.42	Bar	27	17.2	1.06
Pound cake	.07	Cake	26	12.9	1.08
Twix cookie bar	.83	Cookie	20	13.3	1.18
Chocolate bar	.25	2.6 oz	13	9.5	1.26
Doughnut, cake	.34	4 in cake	24	10.9	1.37
Corn chips	.68	Oz	19	10.9	1.44
Soda crackers	3.84	Crckrs	23	11.5	1.48
Angel food cake	.11	Cake	39	22.2	1.53
Snickers bar	.19	3.7 oz	21	11.8	1.56
Wheat thins	11.10	Crckrs	22	13.3	1.61
Whole grain bar, Kudos	1.00	Bar	24	16.0	1.66
Popcorn, air popped	3.22	Cup	26	16.1	1.80
Cupcake with frosting	.76	Cake	33	20.5	1.83
Vanilla wafers	3.57	Cookie	21	14.3	1.98
Graham crackers	3.32	2 1/2 inch sq	23	16.6	2.12
Pretzels	.93	Oz	26	20.5	2.61
Fruit roll ups	1.92	Rolls	27	23.0	3.40

TABLE SIX: GLYCEMIC DENSITY ALPHABETICAL LIST

Note: Dry cereal listings are based upon 1/2 cup of milk for each ounce of cereal.

Food	**GDS**
All Bran, Kellogs, with milk	0.16
Angel food cake	1.53
Apple muffin	0.97
Apples	0.17
Applesauce, no added sugar	0.20
Apricots	0.21
Apricots, dried	0.72
Apricots, in juice	0.27
Asparagus, cooked	-
Avocados	-
Bagel	1.48
Baked beans, canned	0.30
Banana	0.41
Banana bread	1.00
Barley, pearled	0.24
Beets, canned, drained	0.14
Blueberry muffin	0.85
Bran Chex, General Mills, with milk	0.34
Bread, pumpernickel	0.77
Bread, rye	1.08
Bread, seven grain	1.04
Bread, white	1.35

Food	GDS
Bread, whole grain wheat	0.71
Cranberry Sauce	0.86
Cream of Wheat, Nabisco	0.22
Broccoli, cooked	-
Broccoli, raw	-
Brussels sprouts, cooked	-
Buckwheat	0.36
Cabbage, raw, shredded	-
Cantaloupe	0.20
Carrots, raw, chopped	0.05
Cashews	0.06
Cauliflower, cooked	-
Cauliflower, raw	-
Celery, raw, chopped	-
Cheerios, General Mills, with milk	0.39
Cherries, no pits	0.12
Chocolate bar	1.26
Chocolate cake, frosting	0.78
Coco Puffs, General Mills, with milk	0.51
Corn	0.37
Corn bread	1.27
Corn chips	1.44
Corn Flakes with milk	0.55
Corn tortilla	1.02
Cottage Cheese, low fat (1%)	-
Cottage Cheese, nonfat	-
Cottage Cheese, plain	-
Cottage Cheese, reduced fat (2%)	-
Couscous	0.57

Food	GDS
Crispy Rice with milk	0.61
Cucumber, raw, sliced	-
Cupcake with frosting	1.83
Custard	0.18
Dates	1.09
Doughnut, cake	1.37
English muffin	1.49
Fettucinne, egg	0.37
Figs, dried	1.38
Fructose (fruit sugar)	0.76
Fruit cocktail, in juice	0.23
Fruit roll ups	3.40
Golden Grahams, Gen Mills, with milk	0.48
Graham crackers	2.12
Grapefruit	0.10
Grapenuts, Post, with milk	0.47
Grape Nut Flakes, Post, with milk	0.48
Grapes, Thompson, seedless	0.32
Green beans, cooked	-
Green onions, raw	-
Green pepper, raw, chopped	-
Honey	2.20
Ice cream, light, low fat	0.36
Ice cream, nonfat, no sugar added	0.47
Ice cream, premium	0.28
Ice cream, regular	0.41
Jams and jellies	1.43
Kidney beans, canned	0.09
Kiwi	0.25

Food	**GDS**
Lentils, cooked	0.12
Lettuce, iceberg, raw	-
Lettuce, Romaine, raw	-
Lima beans, cooked	0.18
M&M Peanut	0.75
Macaroni & Cheese, boxed	0.65
Mango	0.26
Milk, chcolate (1%)	0.14
Milk, chocolate, nonfat, no sugar added	0.07
Milk, lowfat (1%)	0.06
Milk, nonfat	0.06
Milk, reduced fat (2%)	0.06
Milk, soy	0.03
Milk, whole	0.05
Mini Wheats, frosted, with milk	0.37
Mueslix, Kelloggs, with milk	0.32
Mushrooms, cooked	-
Navy beans, cooked	0.24
Oatmeal cookie with raisins	0.73
Oatmeal, old fashioned	0.14
Oatmeal, one minute	0.21
Olives	-
Oranges	0.17
Pancake, plain	0.78
Peaches	0.14
Peaches, in heavy syrup	0.42
Peaches, in juice	0.15
Peaches, in light syrup	0.23
Peanuts	0.01

Food	**GDS**
Pears	0.19
Pears, in juice	0.16
Peas, cooked	0.21
Pinneapple	0.26
Pinto beans, canned	0.18
Pita	1.14
Pizza, thin, 14 "	0.37
Plumbs	0.16
Popcorn, air popped	1.80
Potato chips	1.00
Potatoes, boiled	0.40
Potatoes, french fried	0.36
Potatoes, mashed milk/butter	0.45
Potatoes, mashed with milk	0.48
Potatoes, Russet, baked	0.63
Pound cake	1.08
Power Bar	1.06
Pretzels	2.61
Prunes	0.61
Pudding	0.33
Pudding, sugar free nonfat milk	0.12
Radishes, raw	-
Raisin Bran, with milk	0.35
Raisins	1.48
Rice, brown	0.47
Rice, white	0.76
Rye crispbread	1.05
Shortbread cookie	0.85
Shredded Wheat, with milk	0.45

Food	**GDS**
Snap peas, raw, whole	-
Snickers bar	1.56
Soda crackers	1.48
Soybeans, fresh	0.05
Spaghetti, whole wheat	0.38
Spaghetti, protein enriched	0.31
Spaghetti, white	0.50
Special K, Kellogg's, with milk	0.40
Spinach, cooked	-
Spinach, raw	-
Squash, winter, cooked	-
Strawberries	0.09
Sweet potato, cooked	0.27
Table sugar	2.44
Tomatoes, cooked	-
Tomatoes, raw, chopped	-
Total, General Mills, with milk	0.45
Twix cookie bar	1.17
Vanilla wafers	1.98
Waffle, plain	0.96
Watermelon	0.21
Wheat thins	1.61
Whole grain bar, Kudos	1.66
Yams, cooked	0.35
Yogurt, fruit on bottom	0.25
Yogurt, nonfat flavored, no sugar added	0.07
Yogurt, nonfat, plain	0.08
Zoom / Wheatena	0.12
Zucchini, cooked	-

PART IV
SUPPLEMENTAL TABLES

TABLE SEVEN: PROTEIN DENSITY TABLES

OUR NEED FOR PROTEIN

Of the three major nutrients protein, carbohydrates and fat, protein is the most important. The body can manufacture carbohydrate and fat from protein, but because protein contains nitrogen, protein cannot be created from fat or carbohydrates. Most nutritional experts recommend two or three daily servings of dairy, one daily serving of nuts or legumes, and one dense serving of lean meat, fish, poultry, low fat cheese or soy protein substitute.

Most protein lists order meat, fish and poultry by fat percentage. This type of measurement, however, leaves out carbohydrates. I prefer to use the reverse procedure of measuring percentage of calories due to protein. Obviously, the desirable thing when consuming our dense protein selection is to eat as high on the percentage scale as possible to cut down on overall calories and saturated fat.

All servings are within or close to the Mayo Clinic 110 calorie recommendation. Exceptions are highlighted. One 3 oz serving of meat is about the size of a deck of cards. One 1 oz serving of cheese is about the size of a pair of dice.

PROTEIN DENSITY OF FOODS

Food	Serv	Cal	Prot	% Prot
Very High Protein 75% to 100%				
Tuna, in water	3 oz	90	21	93
Cod	3 oz	90	21	93
Salmon, Coho	2 oz	80	19	93
Haddock	3 oz	93	21	90
Cheddar cheese, nonfat, shredded	1/2 cup	80	18	90
Crab	3 oz	81	18	89

Food	Serv	Cal	Prot	% Prot
Very High Protein 75% to 100% (Continued)				
Egg Whites	3/4 cup	90	20	87
Buffalo	3 oz	111	24	86
Shrimp	3 oz	84	18	86
Flounder	3 oz	99	21	85
Mozzarella cheese, nonfat, shredded	2 oz	90	18	80
Halibut	2 oz	82	16	78
Scallops	3 oz	93	18	77
Turkey, light, no skin	2 oz	94	18	76
Venison	2 oz	96	18	75
Textured Soy Vegetable Protein	1/2 cup	80	15	75
High Protein 50% to 74%				
Cheddar cheese, low fat	2 oz	100	18	72
Chicken, light, no skin	2 oz	90	16	71
Cottage cheese, nonfat	1/2 cup	80	14	70
Cottage cheese, low fat (1%)	1/2 cup	81	14	69
Cheddar cheese, nonfat, slices	3 slices	90	15	67
Swiss cheese, nonfat, slices	3 slices	90	15	67
Trout	2 oz	86	14	65
Colby cheese, low fat	2 oz	100	16	64
Pork, tenderloin, trimmed	2 oz	114	18	63
Beef liver	2 oz	92	14	61
Ham, canned	3 oz	99	15	60
Bacon, Canadian	3 oz	99	15	60
Tuna, in oil, drained	2 oz	106	16	60

Food	Serv	Cal	Prot	% Prot
High Protein 50% to 74% (Continued				
Turkey, dark, no skin	2 oz	110	16	58
Beef, round, trimmed	1 1/2 oz	93	14	58
Mushrooms, cooked	1 cup	28	4	57
Egg substitute	1/3 cup	70	10	57
Lobster	3 oz	84	12	57
Beef, top sirloin, trimmed	1 1/2 oz	86	12	56
Cottage cheese, reduced fat (2%)	1/2 cup	100	14	56
Pork, loin, trimmed	1 1/2 oz	99	14	55
Lamb chops, trimmed	1 1/2 oz	100	14	54
American cheese, nonfat slices	3 slices	90	12	53
Chicken, dark, no skin	1 1/2 oz	81	10	52
Medium Protein 40% to 49%				
Chicken, light, with skin	1 1/2 oz	86	10	49
Spinach, cooked	1 cup	41	5	48
Sausage, low fat	3 oz	99	12	48
Cottage cheese	1/3 cup	77	9	48
Beef, T-bone, trimmed	1 1/2 oz	87	10	48
Beef, flank, trimmed	1 1/2 oz	100	12	48
Catfish	2 oz	86	10	47
Beef, ground, extra lean	1 1/2 oz	112	12	42
Beef flank, untrimmed	1 1/2 oz	112	12	42
Beef, top sirloin, untrimmed	1 1/2 oz	114	12	42
Chicken, dark, with skin	1 1/2 oz	100	10	42
Yogurt, plain, nonfat	3/4 cup	95	10	41

Food	Serv	Cal	Prot	% Prot
Medium Protein 40% to 49% (Continued)				
Pork, tenderloin, untrimmed	1 1/2 oz	118	12	41
Beef, ground, lean	1 1/2 oz	118	12	41
Asparagus, cooked	1 cup	40	4	40
Milk, nonfat	1 cup	90	9	40
Sardines, in water	2 oz	100	10	40
Salmon, Chinook	1 1/2 oz	105	10	40
Beef rib, trimmed	1 1/2 oz	105	10	40
Beef, round, untrimmed	1 1/2 oz	120	12	40
Mozzarella cheese	1 1/2 oz	120	12	40
Low Protein 30% to 39%				
Beef, ground	1 oz	83	8	39
Soy tempeh	2 oz	91	12	39
Ricotta cheese, part skim	2 oz	86	8	37
Parmesan cheese	3/4 oz	97	9	37
Kielbasa	1 oz	88	8	36
Yogurt, nonfat, fruit flavor, no sugar	1 cup	100	9	36
Soybeans, fresh	1/4 cup	94	8	34
Pork, loin, untrimmed	1 oz	85	7	33
Romano cheese	1 oz	110	9	33
Eggs	1 egg	75	6	32
Beef, T-bone, untrimmed	1 oz	88	7	31
Milk, low fat (1%)	1 cup	102	8	31
Lentils, cooked	1/2 cup	115	9	31
Sardines, in oil, drained	1 oz	77	6	31
Vegetable beef soup	1 cup	78	6	31
Swiss cheese	1 oz	107	8	30
Lima beans, cooked	1/2 cup	108	8	30

Food	Serv	Cal	Prot	% Prot
Very Low Protein 0 to 29%				
Hamburger, large	**1 burg**	**692**	**50**	**29**
Subway sandwich	**1 sand**	**410**	**29**	**28**
Split pea soup	1/2 cup	93	6	28
Chicken noodle soup, chunky	1 cup	114	8	28
Provolone cheese	1 oz	100	7	28
Milk, chocolate, fat free, no sugar added	3/4 cup	98	4	28
Peas, cooked	1/2 cup	116	8	28
Milk, reduced fat (2%)	1 cup	121	8	26
Monterrey Jack cheese	1 oz	106	7	26
Fish fillet, breaded	1 fillet	116	8	26
American cheese	1 oz	93	6	26
Bacon	2 strips	104	7	26
Cheddar cheese	1 oz	110	7	25
Parmesan cheese, nonfat	3 oz	96	6	25
Navy beans, cooked	1/2 cup	128	8	25
Colby cheese	1 oz	112	7	25
Milk, soy	1 cup	113	7	25
Pumpkin seed	1 oz	148	0	24
Egg muffin	**1 muff**	**280**	**17**	**24**
Lamb chops, untrimmed	1 oz	102	6	24
Pinto beans, canned	1/2 cup	103	6	23
Pudding, sugar free in nonfat mil	1/2 cup	70	4	23
Bologna	1 1/2 oz	105	6	23
Kidney beans, canned	1/2 cup	108	6	22
Bacon, turkey	1 oz	90	5	22
Beef, rib, untrimmed	1 oz	109	6	22

Food	Serv	Cal	Prot	% Prot
Very Low Protein 0 to 29% (Continued)				
Pork and Beans, canned	1/2 cup	110	6	22
Taco	**1 taco**	**371**	**20**	**22**
Chicken Noodle soup	1 cup	75	4	21
Milk, whole	3/4 cup	112	6	21
Herring	1 1/2 oz	112	6	21
Fish sticks	1 stick	114	5	21
Clam chowder	3/4 cup	100	4	21
Tomato soup, made with nonfat milk	1 cup	115	6	21
Spaghetti, protein, cooked	1/2 cup	115	6	21
Refried beans, canned	1/3 cup	90	5	21
Milk, chocolate (1%)	3/4 cup	118	6	20
Baked beans, canned	1/2 cup	120	6	20
Hamburger, small	**1 burg**	**265**	**13**	**20**
Minnestone soup	1 cup	82	4	20
Yogurt, fruit on bottom	1/2 cup	115	6	11
Custard	1/2 cup	74	4	19
Sausage	1 oz	128	6	19
Spaghett, whole wheat	1/2 cup	87	4	18
Pizza, thin 14"	**1 piece**	**305**	**14**	**18**
Pepperoni	1 oz	132	6	18
Ricotta cheese	2 oz	82	4	17
Peanuts	15 nuts	83	4	17
Power bar	**1 bar**	**236**	**9**	**15**
Peanut butter	1 tbsp	105	4	15
Tomato soup, made with regular milk	1/2 cup	80	3	15
Almonds	12 nuts	82	3	15

Food	Serv	Cal	Prot	% Prot
Very Low Protein 0 to 29% (Continued)				
Fettuccine, egg, cooked	1/2 cup	110	4	15
Spaghetti, white, cooked	1/2 cup	110	4	15
Sunflower seeds	1/2 oz	83	3	15
Oatmeal, old fashioned	1/2 cup	75	3	13
Frankfurter	1 wien	136	5	13
Ice cream, nonfat, no sugar added	1/2 cup	92	13	13
Burrito	**1 burr**	**224**	**7**	**12**
Pudding	1/2 cup	135	4	12
Chicken nuggets	1 nugg	97	3	11
Cream of chicken soup	1 cup	117	3	10
Cashews	9 nuts	82	2	10
Macaroni & Cheese, boxed	**1/2 cup**	**245**	**6**	**10**
Hazelnuts	2 tbsp	92	2	9
Walnuts	7 halves	93	2	9
Brazil nuts	1/2 oz	93	2	9
Corn bread	1/2 piece	94	2	9
Ice cream, regular	1/2 cup	143	3	8
Rice, brown, cooked	1/2 cup	109	2	7
Ice cream, premium	**1/2 cup**	**188**	**3**	**6**
Pecans	10 halve	100	2	6
Macadamia nuts	5 nuts	102	1	4

VEGETARIAN PROTEIN DENSITY OF FOODS

Food	Serv	Cal	Prot	% Prot
Very High Protein 75% to 100%				
Cheddar cheese, nonfat, shredded	1/2 cup	80	18	90
Egg Whites	3/4 cup	90	20	87
Mozzarella cheese, nonfat, shredded	2 oz	90	18	80
Textured Soy Vegetable Protein	1/2 cup	80	15	75
High Protein 50% to 74%				
Cheddar cheese, low fat	2 oz	100	18	72
Cottage cheese, nonfat	1/2 cup	80	14	70
Cottage cheese, low fat (1%)	1/2 cup	81	14	69
Cheddar cheese, nonfat, slices	3 slices	90	15	67
Swiss cheese, nonfat, slices	3 slices	90	15	67
Colby cheese, low fat	2 oz	100	16	64
Mushrooms, cooked	1 cup	28	4	57
Egg substitute	1/3 cup	70	10	57
Cottage cheese, reduced fat (2%)	1/2 cup	100	14	56
American cheese, nonfat slices	3 slices	90	12	53
Medium Protein 40% to 49%				
Spinach, cooked	1 cup	41	5	48
Cottage cheese	1/3 cup	77	9	48
Yogurt, plain, nonfat	3/4 cup	95	10	41

Food	Serv	Cal	Prot	% Prot
Medium Protein 40% to 49% (Continued)				
Milk, nonfat	1 cup	90	9	40
Asparagus, cooked	1 cup	40	4	40
Mozzarella cheese	1 1/2 oz	120	12	40
Low Protein 30% to 39%				
Soy tempeh	2 oz	91	12	39
Ricotta cheese, part skim	2 oz	86	8	37
Parmesan cheese	3/4 oz	97	9	37
Yogurt, nonfat, fruit flavor, no sugar	1 cup	100	9	36
Soybeans, fresh	1/4 cup	94	8	34
Romano cheese	1 oz	110	9	33
Eggs	1 egg	75	6	32
Milk, low fat (1%)	1 cup	102	8	31
Lentils, cooked	1/2 cup	115	9	31
Vegetable beef soup	1 cup	78	6	31
Swiss cheese	1 oz	107	8	30
Lima beans, cooked	1/2 cup	108	8	30
Split pea soup	1/2 cup	93	6	28
Provolone cheese	1 oz	100	7	28
Milk, chocolate, fat free, no sugar added	3/4 cup	98	4	28
Peas, cooked	1/2 cup	116	8	28
Milk, reduced fat (2%)	1 cup	121	8	26
Monterrey Jack cheese	1 oz	106	7	26
Very Low Protein 0 to 29%				
American cheese	1 oz	93	6	26
Cheddar cheese	1 oz	110	7	25
Parmesan cheese, nonfat	3 oz	96	6	25

Food	**Serv**	**Cal**	**Prot**	**% Prot**
Very Low Protein 0 to 29% (Continued)				
Parmesan cheese, nonfat	3 oz	96	6	25
Navy beans, cooked	1/2 cup	128	8	25
Colby cheese	1 oz	112	7	25
Pumpkin seed	1 oz	148	0	24
Egg muffin	**1 muff**	**280**	**17**	**24**
Pinto beans, canned	1/2 cup	103	6	23
Pudding, sugar free in nonfat mil	1/2 cup	70	4	23
Kidney beans, canned	1/2 cup	108	6	22
Milk, whole	3/4 cup	112	6	21
Tomato soup, made with nonfat milk	1 cup	115	6	21
Spaghetti, protein, cooked	1/2 cup	115	6	21
Refried beans, canned	1/3 cup	90	5	21
Milk, chocolate (1%)	3/4 cup	118	6	20
Baked beans, canned	1/2 cup	120	6	20
Minnestrone soup	1 cup	82	4	20
Spaghetti, whole wheat	1/2 cup	87	4	18
Yogurt, fruit on bottom	1/2 cup	115	6	11
Custard	1/2 cup	74	4	19
Ricotta cheese	2 oz	82	4	17
Peanuts	15 nuts	83	4	17
Power bar	**1 bar**	**236**	**9**	**15**
Peanut butter	1 tbsp	105	4	15
Tomato soup, made with regular milk	1/2 cup	80	3	15
Fettuccine, egg, cooked	1/2 cup	110	4	15
Spaghetti, white, cooked	1/2 cup	110	4	15

Food	Serv	Cal	Prot	% Prot
Very Low Protein 0 to 29% (Continued)				
Almonds	12 nuts	82	3	15
Sunflower seeds	1/2 oz	83	3	15
Oatmeal, old fashioned	1/2 cup	75	3	13
Ice cream, nonfat, no sugar added	1/2 cup	92	13	13
Burrito	**1 burr**	**224**	**7**	**12**
Pudding	1/2 cup	135	4	12
Cashews	9 nuts	82	2	10
Macaroni & Cheese, boxed	**1/2 cup**	**245**	**6**	**10**
Hazelnuts	2 tbsp	92	2	9
Walnuts	7 halves	93	2	9
Brazil nuts	1/2 oz	93	2	9
Corn bread	1/2 piece	94	2	9
Ice cream, regular	1/2 cup	143	3	8
Rice, brown, cooked	1/2 cup	109	2	7
Ice cream, premium	**1/2 cup**	**188**	**3**	**6**
Pecans	10 halve	100	2	6
Macadamia nuts	5 nuts	102	1	4

TABLE EIGHT: SUPER FOODS

The following foods are considered the most important sources of vitamins, mineral and phytonutrients by leading experts. They provide powerful benefits for cardiovascular health and prevening cancer. I mentioned some of these in my first book. Since then, I have gathered more information. This is a more complete list.

Vegetables

Best Food	Other Foods in Same Class
Broccoli	Brussels sprouts, cabbage, kale, turnips, cauliflower, collards, bok choy, mustard greens, Swiss chard.
Onions	Garlic, scallions, shallots, leeks, chives
Pumpkin	Carrots, winter squash, sweet potatoes, orange bell peppers.
Spinach	Kale, collards, Swiss chard, mustard greens, turnip greens, bok choy, Romaine lettuce, orange bell peppers.

Fruits

Best Food	Other Foods in Same Class
Apples	Pears
Blueberries	Purple grapes, cranberries, boysenberries, raspberries, strawberries, currants, blackberries, cherries, raisins.
Cantaloupe	Honeydew
Dried Super Fruits	Raisins, dates, prunes, other dried berries.

Fruits (Continued)

Best Food	Other Foods in Same Class
Kiwi	Pineapple, guava
Oranges	Lemons, grapefruit, tangerines, limes.
Pomegranates	Plumbs
Tomatoes (the more processed the better)	Red watermelon, pink grapefruit, Japanese persimmons, red fleshed papaya, strawberry guava.
Bananas	Bananas are not usually listed as a superfood because they have more calories than most fruits. However, as a quick energy boost, a banana will give you all the energy you need with a significant amount of nutritional value.

Grains

Best Food	Other Foods in Same Class
Oats	Wheat germ, ground flaxseed, brown rice, barley, whole grain wheat, buckwheat, rye, millet, bulgur wheat, amaranth, quinoa, triticale, kamut, yellow corn, wild rice, spelt, couscous.

Dairy

Best Food	Other Foods in Same Class
Yogurt	Kefir, soy yogurt.

Vegetable Protein

Best Food	Other Foods in Same Class
Beans	Peas, green beans, lentils, snap peas.
Soy	Tofu, soy milk, soy nuts, adamame, tempeh, miso.

Animal Protein

Best Food	Other Foods in Same Class
Wild Salmon	Alaskan halibut, canned albacore tuna, sardines, herring, trout, sea bass, oysters, clams.
Turkey (skinless breast)	Chicken (skinless breast).
Other meats	Lean beef, pork or buffalo on occasion.

Unsaturated Fats

Best Food	Other Foods in Same Class
Avocado	Asparagus, artichokes, olive oil
Extra Virgin Olive Oil	Canola oil
Walnuts	Almonds, pistachios, sesame seeds, peanuts, pumpkin and sunflower seeds, macadamia nuts, pecans, hazelnuts, cashews.

Others

Best Food	Other Foods in Same Class
Cinnamon	A non-calorie source of good health, especially for diabetics.
Dark Honey	An excellent source of polyphenols.
Dark Chocolate	An excellent source of polyphenols.
Garlic	Scallions, shallots, leeks, onions.
Tea (both green and black)	Fluoride = canned fish with bones, toothpaste, water. Flavonoids = cocoa, dark chocolate, red apples, other fresh fruits and vegetables.

REFERENCES

BIBLIOGRAPHY

DIABETES MANAGEMENT

Atkins, Robert C., M.D., Eberstein, Jacqueline A., R.N, Vernon, Mary C., M.D., C.M.D. Atkins Diabetes Revolution. New York: William Morrow Imprints, Harper-Collins Publishers, Inc., 2004.

Barnard, Neal D. M.D. Dr. Neal Barnard's Program for Reversing Diabetes. New York: Rodale Inc., 2007.

Bernstein, Richard K., M.D. Dr. Bernstein's Diabetes Solution. New York: Little, Brown and Company, 2007.

Biermann, June and Toohey, Barbara. The Diabetics Book. New York: G. P. Putnam's Sons, 1994.

Holler, Harold J. R.D., C.D.E., & others, ed. Exchange Lists for Meal Planning. American Diabetes Association and American Dietetic Association, 1995.

ENERGY / CALORIE DENSITY

Bailey, Covert. The Fit or Fat Target Diet. Boston: Houghton Mifflin Company, 1984.

Bailey, Covert and Gates, Ronda. Smart Eating. Boston: Houghton Mifflin Company, 1996.

Govindji, Azmina and Puddefoot, Nina. The Simple 0-to-10 GI Diet. Berkeley, California: Ulysses Press, 2005.

Hunsrud, Donald D. M.D., ed. Mayo Clinic on Healthy Weight. Rochester, Minn: Mayo Clinic Health Information, 2001.

Hunsrud, Donald D. M.D., ed. The Mayo Clinic Plan. New York, Time Inc., 2006.

Rolls, Barbara, Ph.D., The Volumetrics Eating Plan. New York: HarperCollins, 2005.

Rolls, Barbara, Ph.D.and Barnett, Robert A., The Volumetrics Weight-Control Plan. New York: HarperTorch imprint of HarperCollins, 2003.

EXERCISE

Bailey, Covert. Fit or Fat. Boston: Houghton Mifflin Company, 1978.

Bailey, Covert. The Ultimate Fit or Fat. Boston: Houghton Mifflin Company, 1999.

Haung, Alfred. The Complete Tai-Chi: The Definitive Guide to Physical and Emotional Self Improvement. Boston: Charles E. Tuttle Co. 1993.

Hittleman, Richard. Yoga: 28 Day Exercise Plan. New York: Workman Publishing Co. Inc. 1973.

Kit, Wong Kiew. Chi Kung for Health and Vitality: A Practical Approach to the Art of Energy. New York: Random House. 2001.

GLYCEMIC INDEX

Brand-Miller, Jennie, Ph.D. Glycemic Index Values. 2005, University of Sidney. 22 June 2007 <www.glycemicindex.com>.

Brand-Miller, Jennie, Ph.D., Burani, Johanna, M.S., R.D., C.D.E. and Foster-Powell, Kaye, M. Nutr & Diet. The Glucose Revolution Life Plan. New York: Marlow & Company, 2001.

Brand-Miller, Jennie, Ph.D., Burani, Johanna, M.S., R.D., C.D.E. and Foster-Powell, Kaye, M. Nutr & Diet. The New Glucose Revolution Pocket Guide to The Top 100 Low GI Foods. New York: Marlow & Company, 2001.

Brand-Miller, Jennie, Ph.D., Foster-Powell, Kaye, M. Nutr & Diet with Sandall, Philippa. The New Glucose Revolution Low GI Eating Made Easy. New York: Marlow & Company, 2005.

Brand-Miller, Jennie, Ph.D., Wolever, Thomas, M.S., M.D., Ph.D., Foster-Powell, Kaye, M. Nutr & Diet, and Colagiuri, Stephen, M.D. The New Glucose Revolution, The Authoritative Guide to the Glycemic Index. New York: Marlow & Company, 2003.

Brand-Miller, Jennie, Ph.D., Foster-Powell, Kay, M. Nutr & Diet. The New Glucose Revolution Pocket Guide to Sugar and Energy. New York: Marlow & Company, 2004.

Gallop, Rick, The G.I. (Glycemic Index) Diet. New York: Workman Publishing, 2002.

Govindji, Azmina and Puddefoot, Nina. The Simple 0-to-10 GI Diet. Berkeley, California: Ulysses Press, 2005.

Hart, Cheryle R., M.D. and Grossman, Mary Kay, R.D. The Insulin Resistance Diet. New York: McGraw Hill, 2001,

Holford, Patrick. The Holford Low GL Diet. New York: Atria Books, 2004.

LOW CARB DIETING

Atkins, Robert C., M.D. Atkins for Life. New York: St. Martin's Press, 2003.

Atkins, Robert C., M.D. Dr. Atkins' New Diet Revolution. New York: Quill Imprints, Harper-Collins Publishers, Inc., 2002.

Atkins, Robert C., M.D., Eberstein, Jacqueline A., R.N, Vernon, Mary C., M.D., C.M.D. Atkins Diabetes Revolution. New York: William Morrow Imprints, Harper-Collins Publishers, Inc., 2004.

Bernstein, Richard K., M.D. Dr. Bernstein's Diabetes Solution. New York: Little, Brown and Company, 2007.

Carpender, Dana. How I Gave Up My Low-Fat Diet and Lost 40 Pounds. Glouchester, Mass: Fair Winds Press, 2003.

Heller, Richard and Heller, Rachel. The Carbohydrate Addict's Lifespan Program. New York, A Plume Group imprint of Penguin Group, 1997.

McDonald, Lyle. The Ketogenic Diet. Austin, Texas, Lyle McDonald, 1998.

Graff, Cynthia Stamper with Jerry Holderman. Lean for Life - Phase One Weight Loss. Torrance, California: Griffin Publishing. 2001.

MODERATE CARB DIETING

Agatston, Arthur, M.D. The South Beach Diet. New York: Rodale, Inc., 2003.

Sears, Barry, Ph.D. with Lawren, Bill. The Zone. New York, HarperCollins, 1995.

Sears, Barry, Ph.D. A Week in the Zone. New York, HarperCollins, 2000.

Stewart, H. Leighton, Bethea, Morrison C., M.D., Andrews, Sam S., M.D., and Balart, Luis A. M.D. The New Sugar Busters, New York, Ballentine Books, 2003.

OTHER NUTRITION

Buettner, Dan, "The Secrets of Long Life" ,National Geographic November, 2005.

Duyff, Roberta Larson, M.S, R.D., C.F.C.S. The American Dietetic Association's Complete Food and Nutrition Guide. New York: John Wiley & Sons, Inc., 1998.

Heller, Marla, M.S., R.D. The Dash Diet Action Plan. Northbrook, Illionois, Amidon Press, 2005.

Pratt, Steven, M.D. and Matthews, Kathy. SuperFoods Rx. New York: William Morrow Imprint, HarperCollins, 2004.

Pratt, Steven, M.D. and Matthews, Kathy. SuperFoods HealthStyle. New York: William Morrow Imprint, HarperCollins, 2006.

Rinzler, Carol Ann, M.A. Nutrition for Dummies, 2nd Ed. New York: Wiley Publishing, Inc., 1999.

Sheps, Sheldon G. M.D., ed. Mayo Clinic on High Blood Preasure. Rochester, Minn: Mayo Clinic Health Information, 2002.

Vartabedian, Dr. Roy and Matthews, Kathy, Nutripoints. Carlsbad, California: Designs for Wellness Press, 2005.

Willett, Walter C., M.D. & Harvard School of Public Health. Eat, Drink and Be Healthy. New York: Simon & Schuster, 2001.

REFERENCE

Blonz, Ed, Dr. The Nutrition Doctor's A-to-Z Food Counter. New York: Signet Division, New American Library, Penguin Putnam Inc., 1999.

Natow, Annette B., Ph.D., R.D., and Heslin, Jo-Ann, M.A., R.D. The Complete Food Counter, New York: Pocket Books Divison of Simon & Schuster, Inc., 2003.

Search the National Nutrient Database For Standard Reference. 2007, USDA Agricultural Research Service. 22 June 2007 <www.nal.usda.gov/fnic/foodcomp/search>.

SEROTONIN MANAGEMENT

DesMasons, Kathleen, Ph. D. Potatoes Not Prozac. New York: Simon and Schuster Paperbacks, 1998.

Hart, Carol, Ph.D. Secrets of Serotonin. New York: St. Martin's Press, 1996.

Hart, Cheryle R., M.D. and Grossman, Mary Kay, R.D. The Feel Good Diet. New York: McGraw Hill, 2007,

Wurtman, Judity J. Ph.D. The Serotonin Solution. New York: Fawcett Columbine, 1996.

VEGETARIANISM

Barnard, Neal D. M.D. Dr. Neal Barnard's Program for Reversing Diabetes. New York: Rodale Inc., 2007.

Diamond, Harvey. Fit for Life, Not Fat for Life. Deerfield Beach, Florida: Health Communications, Inc., 2003

Havala, Suzanne, M.S., R.D. Being Vegetarian for Dummies. Hoboken, New Jersey, Wiley Publishing, 2001.

Other Books by Richard A. Price

How I lost 80 Pounds with Smart Carb Eating
Secrets of Effective Weight Management
Published by Infinity Publishing

Rich can be contacted at:

richardalanprice@gmail.com

Made in the USA
Monee, IL
08 November 2026

67377466R00077